n|p|k indus

trial design

In 1991 the director of the 'Stedelijk Museum' in Amsterdam invited npk industrial design to exhibit a selection of works from what by then had already become an impressive oeuvre. npk industrial design accepted the invitation with some hesitation, for they felt that a public showing was not without its risks, and certainly at a place like the 'Stedelijk'. The 'Function Forms Style' exhibition and its attendant catalogue would, after all, be hallmarks against which to measure the further development of the agency. Presentation in a museum would also place their work in a very specific context of art and uniqueness, one which in fact they had always avoided. They did not and do not want to act as solo artists, but as equal partners with their industrial clients.

Much has happened since then, and now npk industrial design itself is presenting a public showing in Vision & Precision, a book featuring a selection of work from the past ten years. After having arrived at that first presentation – so admired at the time by both the public and professionals – how much room would be left for the further development and broadening of npk industrial design's mentality and working method?

npk industrial design was founded in 1978 and has grown from a small agency run by Peter Krouwel and Wolfram Peters into a business that now boasts five partners and sixty highly qualified staff members. After the third founding partner, Bruno Ninaber van Eyben, left in 1997, Jan Witte and Jos Oberdorf joined the Dutch branch and Detlef Rhein the German branch of the agency as new partners.

From the very beginning npk industrial design has focused on the design of high-quality technical and functional products for use in people's living and working quarters and in what is called the public space, such as the street and public transportation. These are products produced in large series on an industrial scale, or of which only a few examples are manufactured. Some designs are as small as a microchip while others are extremely large, such as the tramline to IJburg, the new Amsterdam city sector built on land reclaimed from the waters of the IJsselmeer.

Micro to macro professionalism

npk industrial design has always consciously chosen to be capable of taking on everything from micro to macro. This however, is not so simple. Small products compel a designer to employ the precision and detailing of a jewellery maker. Large-scale designs demand the power of the large gesture in order to maintain unity and clarity of visual expression. Working on large-scale products teaches that visual emphasis on millimetres is detrimental to a view of the whole, whereas small-scale products demonstrate that the large gesture must be carried through right down to the details. One purpose of this alternation of scale is to force the designers of npk industrial design to regularly be able to see things in a different light. But it also entails a wide-ranging professionalism and, most assuredly, the pure enjoyment of the professional challenge.

In terms of technical production the differences between projects are just as great, so that the agency must have an equally broad range of technical expertise at its command. npk industrial design goes a long way in this. The model-making workshop has developed into a place with the air of a laboratory, where products can be made and tested down to the minutest detail. The prototypes and first production runs used to test a design and prepare for its production are made here by hand or with computer added tools. Specialised teams oversee the outside manufacture of moulds and other tools necessary for producing the designs. This means that npk industrial design maintains control over the quality of every aspect of the product up until the very last moment.

The idea behind this broad-ranging micro-to-macro scale and the desire to maintain control over all technical aspects is that working on different levels of complexity, functionality, scale and technology is inspiring and stimulates involvement and interaction. It repeatedly forces designers to search for designs that are in balance with the size and function of the product. This breadth of scale is a striking characteristic of npk industrial design. It might even be questioned whether any other design agency could be found with an equal experience and knowledge of dealing with such a variety of commissions so professionally and so consistently.

npk industrial design regularly designs new products of which there are few or no existing examples and for which a lot still has to be worked out. These are always complex products incorporating new knowledge that cannot be dreamt up in an afternoon, but are the result of extensive research and experimentation.

The demands that such designs must meet, whether they be large or minuscule, make it necessary to work systematically in order not to go astray. npk industrial design uses a time-tested method for this, in which a design is put through a series of stages that begins with research and analysis of the problem and that gradually becomes more concrete in sketches and models. At the heart of this process is the stage in which all conditions are correlated as optimally as possible and condensed into what is a design in its most elementary and descriptive form: a material example that can be industrially reproduced.

But what is at issue here is not just the technical/industrial manufacturability or the functionality of a design, however important these may be. The form certainly is just as important an aspect, perhaps even the most important one. The form is not only what we see first, it also is the material vehicle for all functional properties. The visual expression of a product also contains a cultural dimension that gives it a unique place in time. This seems to be an autonomous process that goes further than the function and manufacturability of a product. Moreover, it hardly appears possible to influence it rationally, for even the most sober designers are ultimately children of their time, bound to it by their personal experience of the world around them.

The greatest strength of npk industrial design is its capacity to realise a clear integration of the three most important aspects of a product: industrial manufacturability, functionality and form, which includes the cultural dimension. This is a difficult and time-consuming process of critically searching and trying and rejecting and starting afresh and continuing until everything falls into place.

Clients are understandably in a hurry and are often surprised and perhaps even impatient about this stubborn plodding on until the last millimetre, which sometimes seems very far away indeed. The fact that npk industrial design has successfully managed for 25 years to persevere in this stubborn enthusiasm in the presence of its clients shows that the results are convincing. On the international front as well, there are very few agencies where designs are made so clinically and systematically yet at the same time so intensely and enthusiastically, and where self-criticism is so strong and kept up for so long. The desire for perfection and the urge to get everything possible out of a design means that npk industrial design occasionally steps outside the specifications of a commission at its own risk and initiative in order to see if all possibilities have been sought out, even if this is just for its own satisfaction. Sometimes the limits have already been reached and there are no more new steps to be taken. It often happens, however, that npk industrial design surprises a client with a new idea which can give the product an added dimension of quality.

Development from strength to strength

Since the exhibition in the Stedelijk Museum, npk industrial design has grown from strength to strength. One logical development is that the agency has branched out into other fields, such as graphic design. Many products involve graphic elements in the form of brand names and logos, symbols indicating how to operate them, and of course manuals are also necessary. In many instances, the interaction between user and product occurs through a user interface. A graphic design department was set up to cover all these aspects. It now has a momentum of its own, which extends further than simply providing support for the product designers. For special projects, however, the department works intensively with professionals outside the consultancy, such as with typographer Gerard Unger for the new nation wide signs for the ANWB automobile association.

The personal interest of one of the partners in products for the public space has led to a new line of products that includes benches, streetlamps and signposting systems. So much knowledge and experience has been acquired in this area by now that npk industrial design is capable of handling entirely by itself, the entire equipment of tramlines, complete with tram stops and lighting systems. For the new High Speed Line (HSL), npk industrial design not only designed the vertical components but also conceived the track and its methods of production and implementation. According to their specific way of working, this will be an industrially produced, modular product in concrete. Traditionally, a civil engineering agency would have been asked to do this.

npk industrial design has spread its wings internationally. For several years now, it has operated an office in Hamburg so as to be both logistically and culturally closer to German clients who had already found their way to the agency. Plans are currently underway for a new branch in Boston, USA.

In addition to being an account of a successful agency, this book reveals the emotional interest that npk industrial design has in the medium of photography. This goes much further than the documentation of its own designs. From the very first, the consultancy displayed interest by maintaining its own professional photo studio and working with outside photographers such as Gerrit Schreurs. His photographs depict the designs not just through his own eyes, but are the result of an intense dialogue with the agency. At first sight they seem businesslike and cool – as does npk industrial design on first acquaintance. Upon closer inspection, however, one notices how the expression of the material and a refined use of shadow accentuates the personal intentions of the designer. For photography done in its own studio, npk industrial design did work with Gerrit Schreurs to develop a visual concept that imparts unity to all its presentations.

Structural set-up of Vision & Precision

In a surprising and direct manner, Vision & Precision provides insight into the deeper emotional considerations that determine a design, and elegantly solves two fundamental problems as well, that always arise when we look at illustrations of utilitarian objects. A picture of a product is in fact a stripped-down representation of reality. First of all, the size is not right and the perspective in the two-dimensional image disrupts our estimation of scale. The backside is not visible and there is no movement of any kind, whereas many of our instruments and appliances must move in order to function. Our perception of, and thus our judgement about, many of these products, whether it be a baby's seat, a chair or a laboratory instrument, is largely determined by tactile experience, and this is missing when we look at a picture. Aspects such as weight and balance, the texture and the temperature of the surface, the resistance, the sound of knobs and handles, the changing reflections of light on the surface, are all things that cannot be experienced through a picture. Above all, is missing the context in which the illustrated products function, because our utilitarian objects always operate in an environment with other products.

Another problem presents itself when looking at pictures of products. In the 20th century our serialised consumer goods acquired the dual role of performing a function and expressing the personality of the owner. They play a role in our lives as sober and trusty fulfillers of functions, carrying out tasks to serve our comfort and safety. But in addition to that, design has been given a remarkably dominant role in the nonmaterial expression of our self-image or the image we would like to present of ourselves. It looks as if users have relinquished part of their own expression of their self-image - to their material surroundings. 'Show me your consumer goods and I will tell you who you are,' is no idle remark. Apparently, different individually determined interpretations can be attached to the same product. Consumer goods thus become compliantly bearers of a multitude of meanings.

npk industrial design has come up with a surprisingly effective way of making these two problems of perception discernible and manageable for the user of this book. Each illustration of a design in this book has an illustration next to it as inseparable counterpart. The products are pictured in isolation, without context, but the counterpart illustration does the work. These images - usually photos - conjure up all sorts of personal interpretations that colour the viewer's perception of the picture of the product. These counterpart images make it clear straightaway that everybody looks at the products around them in their own way and with their own knowledge and insight, and particularly with their own perception of their environment. It also becomes clear how differently our utilitarian objects can be interpreted and how little designers can do to steer this. They cannot determine how users should look at things, because there simply is no unequivocal way of looking. Their strength as designers must be to offer a design that supports ambiguous interpretations without losing its character.

The use of counterpart images also partially eliminates the problem of an incomplete representation of reality. This phenomenon is no longer so bothersome because our spontaneous associations evoke a reality - which perhaps could be called virtual - in which the incomplete image is also capable of evoking a reality, which moreover is loaded with emotion.

As such, the book not only elucidates what npk industrial design does. In a visually captivating manner, it also makes the viewer aware of the many and extremely varied relations we have with the hundreds of different functional products in our daily lives. Vision & Precision is certainly more than just a picture book. Studying it not only yields an interesting account but also an exciting interaction between image and perceiver, who becomes aware that other associations can arise within a month or a year because new experiences will have been acquired and additional knowledge gained.

Reyer Kras

Reyer Kras (1940) studied industrial design and art history. He taught at Delft University of Technology (industrial design) and at the University of Leiden (design history). From 1982 to 2001, he worked as the curator for industrial design at the Stedelijk Museum in Amsterdam. In 1996, he founded the office Aurelius Industrial Design for product development and museological projects. He is the author of many publications and teaches regularly both at home and abroad.

clean

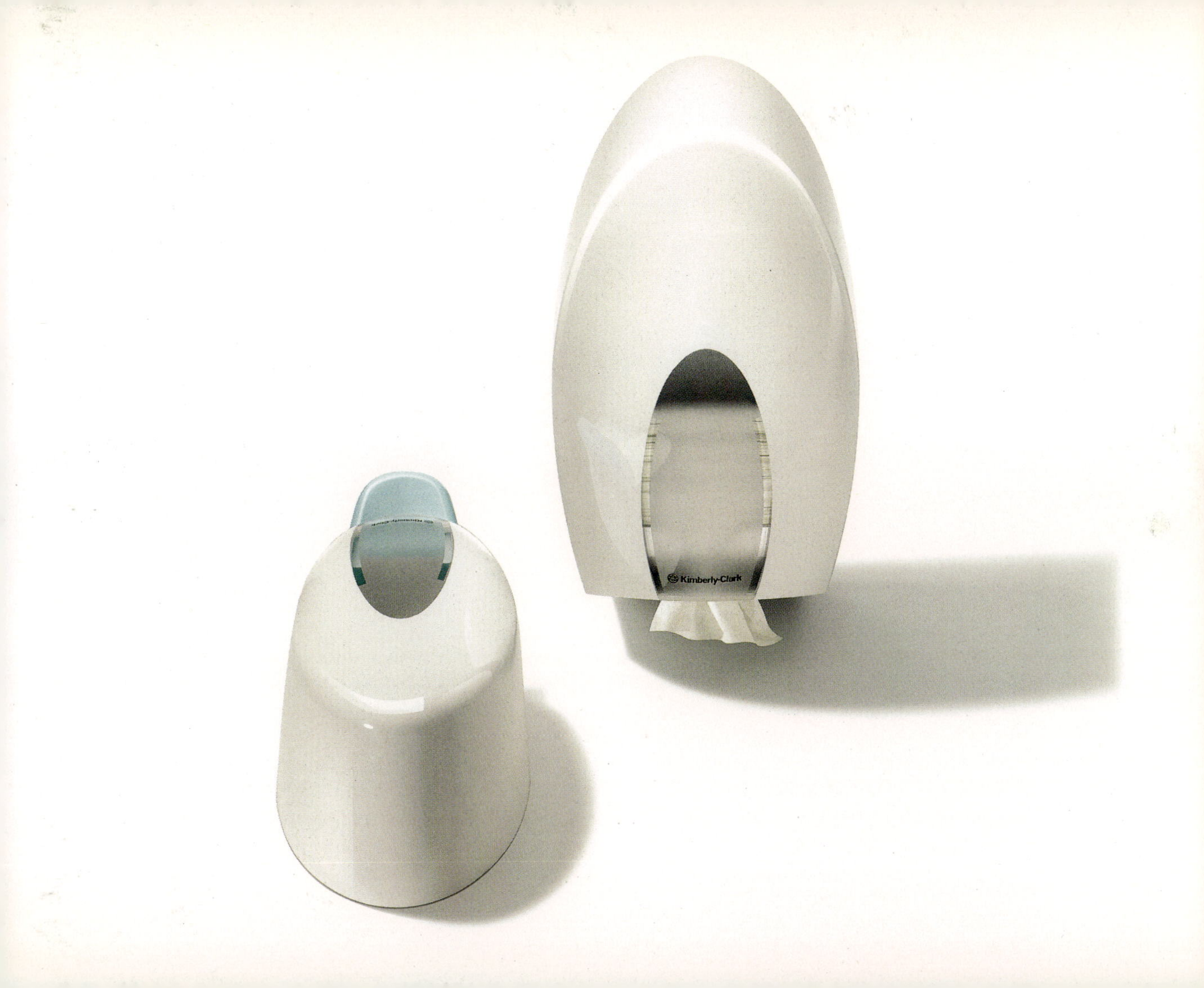
Kimberly-Clark

XL 850

Kimberly-Clark

ASM

cp future
START GC
H_2-GENERATOR

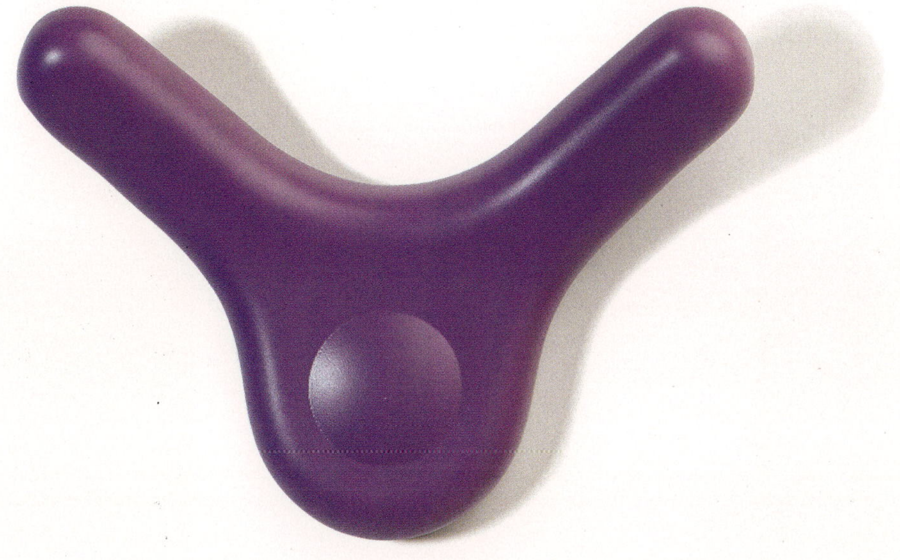

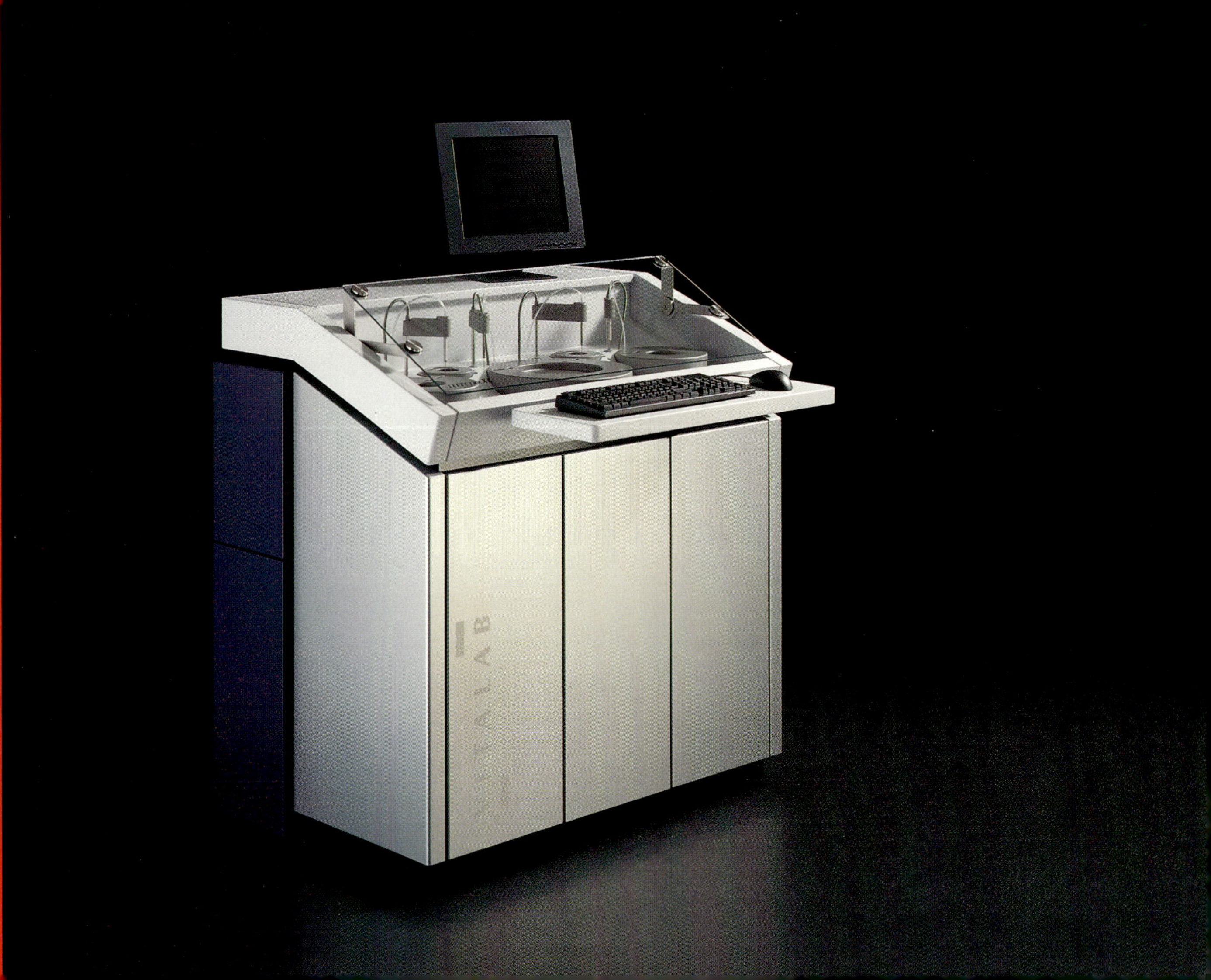
VITALAB

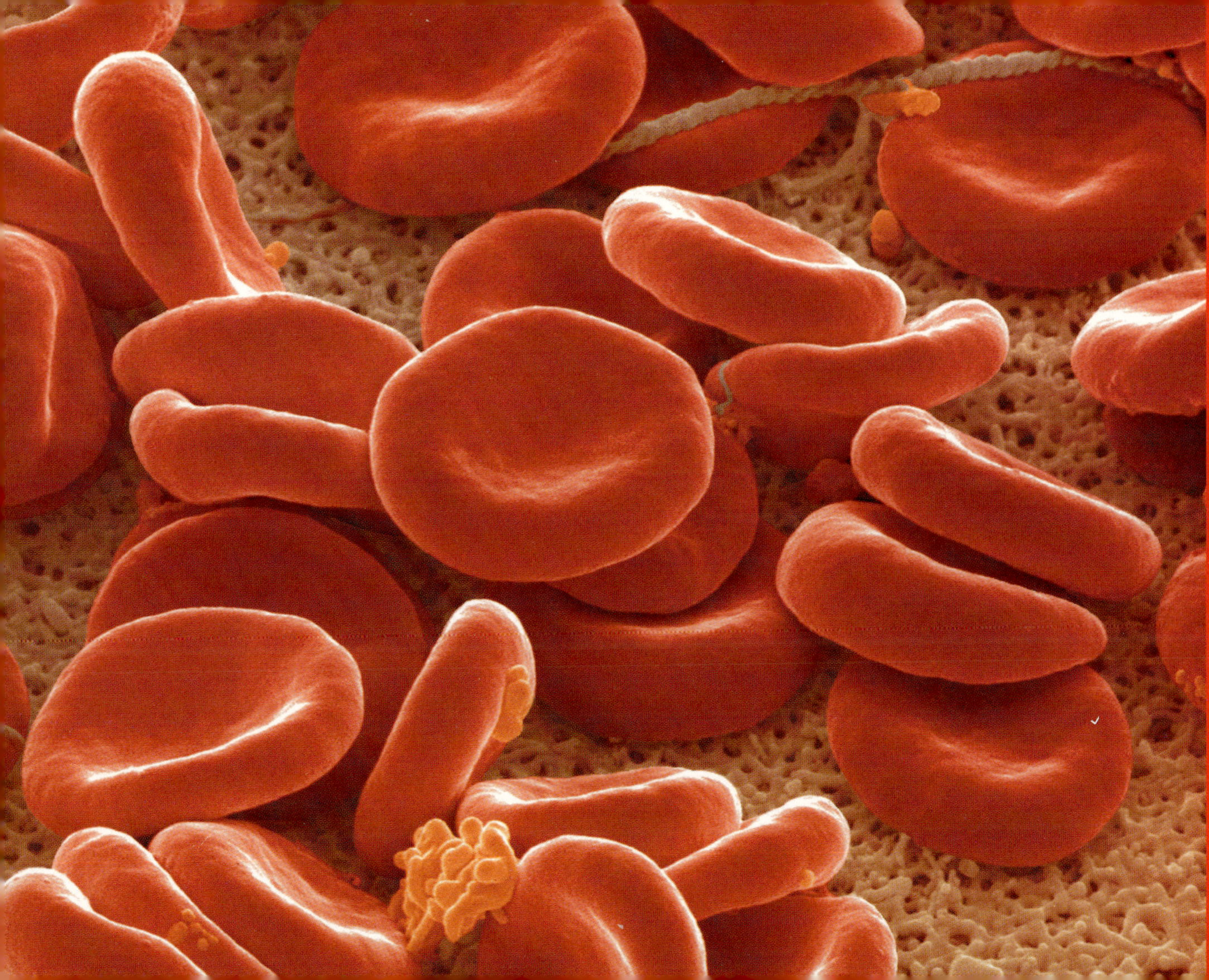

go
FLY DELTA
Butler St
Freedom Parkway
Carter Center
Williams St
Downtown
Ga. Dome
EXIT 99

Leiden -
Leiderdorp 2

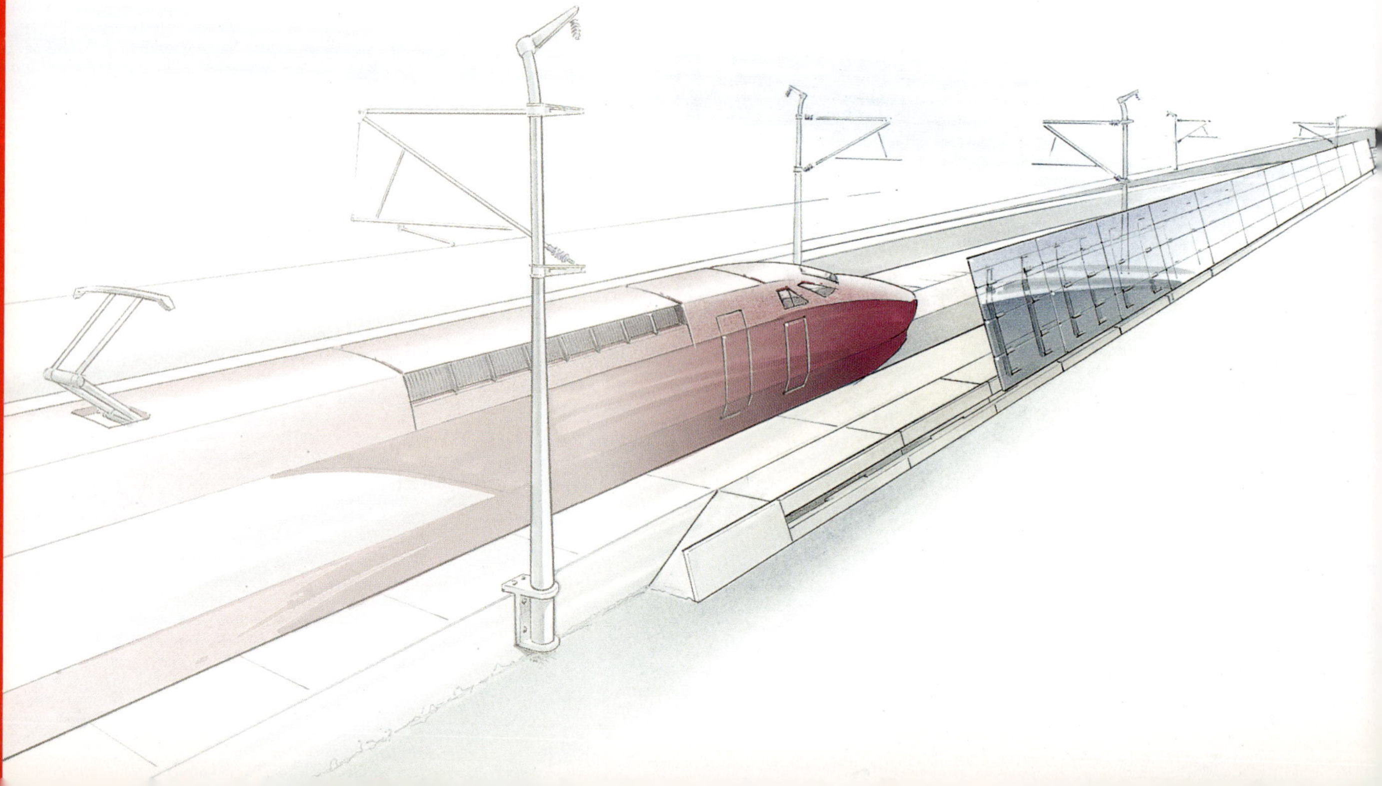

San Giovanni in Laterano
Santa Croce in Gerusalemme
Colosseo
San Salvatore
Pantheon
San Pietro
San Pietro
Colosseo
Pantheon

RGX 40001 AS
ASML
High Technology from Holland
HOISTING POINT
HOISTING POINT
ASML
www.asml.com
Frigotainer
FORKLIFT

roeiersvlet

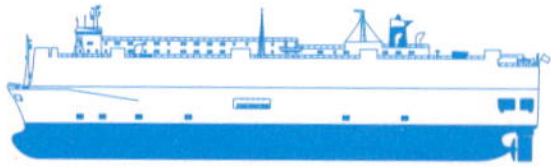
autocarrier

bulkcarrier

baggerschip

containerschip

containerbinnenvaartschip

havendienstboot

loodstender

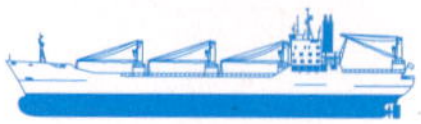
koelschip

duwbakcombinatie

bunkerschip

sleepboot

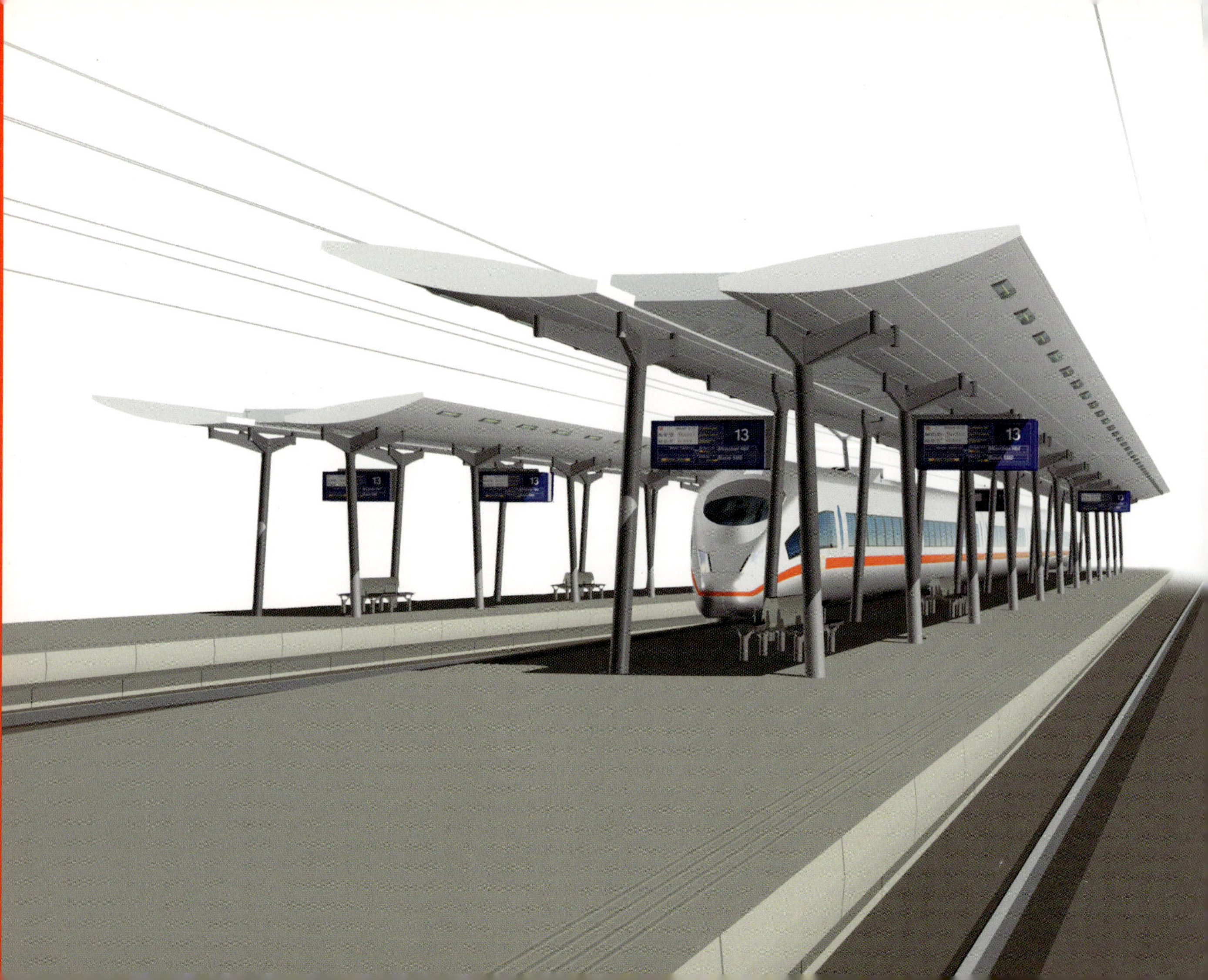
13

2602
২৬০২

work

DS 202
NEG
M2
NEG 03/176
III
634 0.124
2747

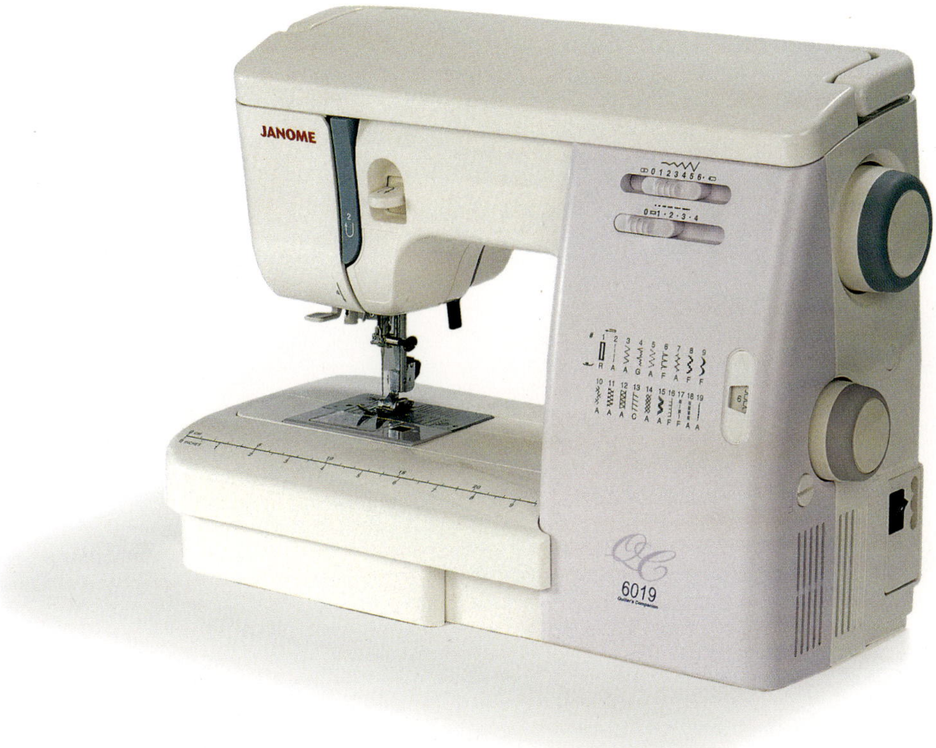
JANOME
6019

A
B
F
G
C
E
D
E04
37
Beveiliging beheer en ondersteuning
Security control and support
P. de Groot
J. de Lahaye-Beckers

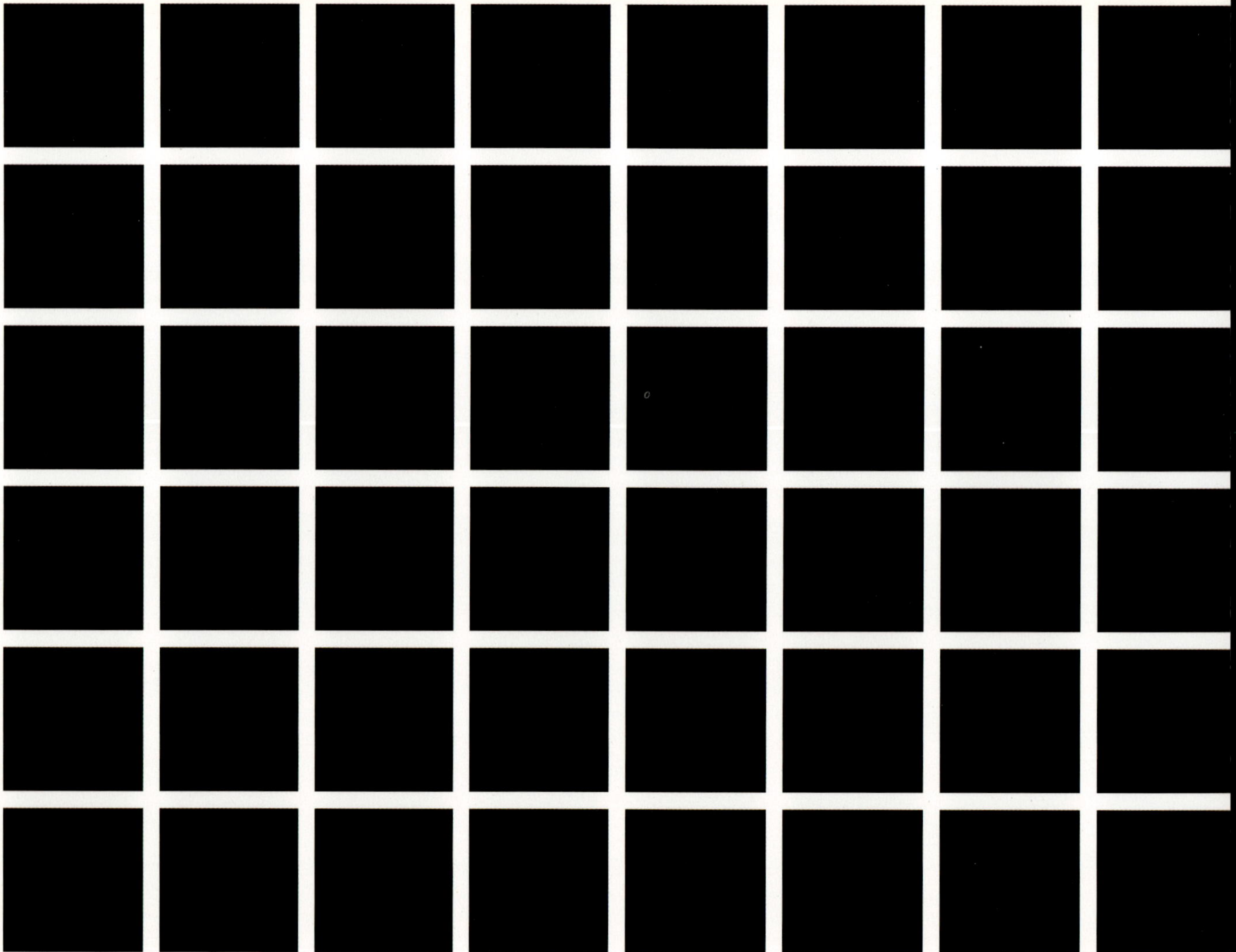

randstad
randstad

eat

Berkel

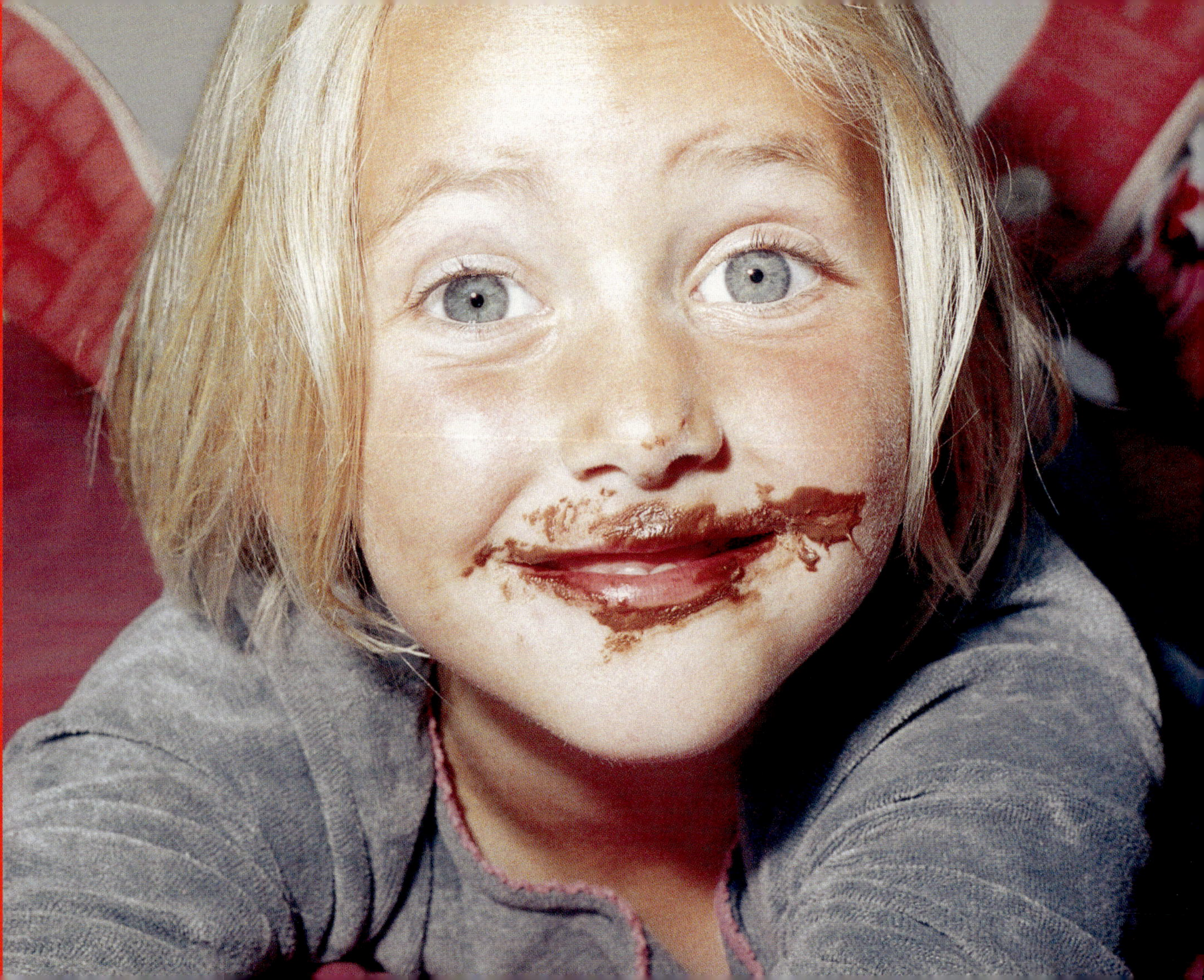

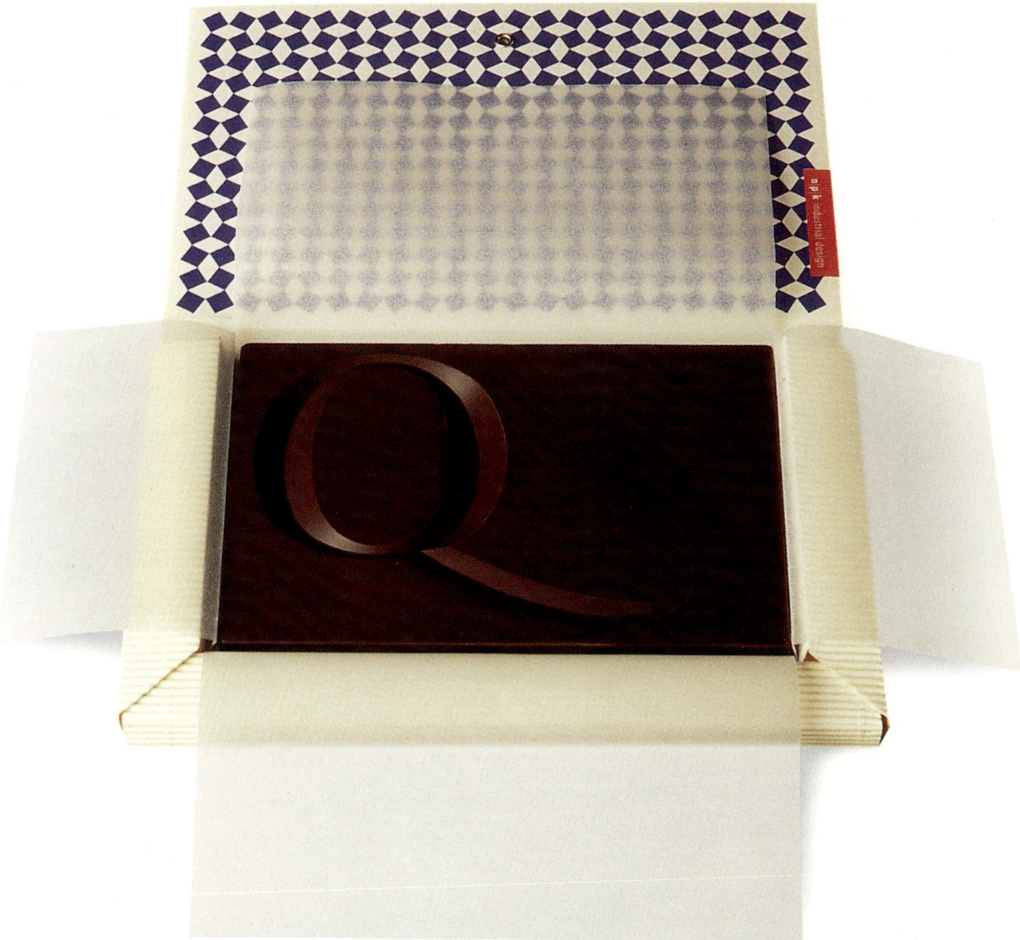
n|p|k industrial design

de koffie is klaar

SOLVAY
SOLVAY PHARMA

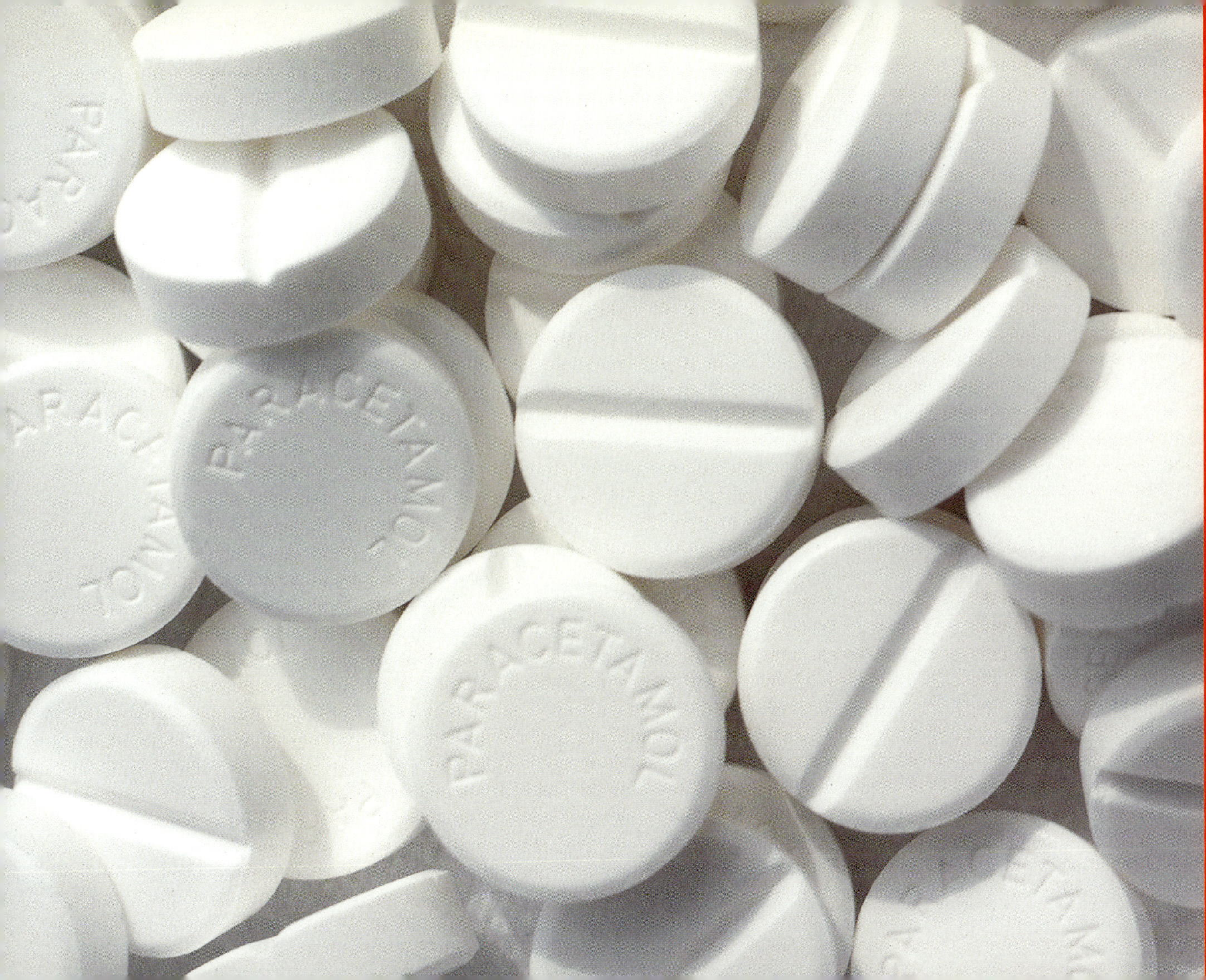
PARACETAMOL
PARACETAMOL
PARACETAMOL

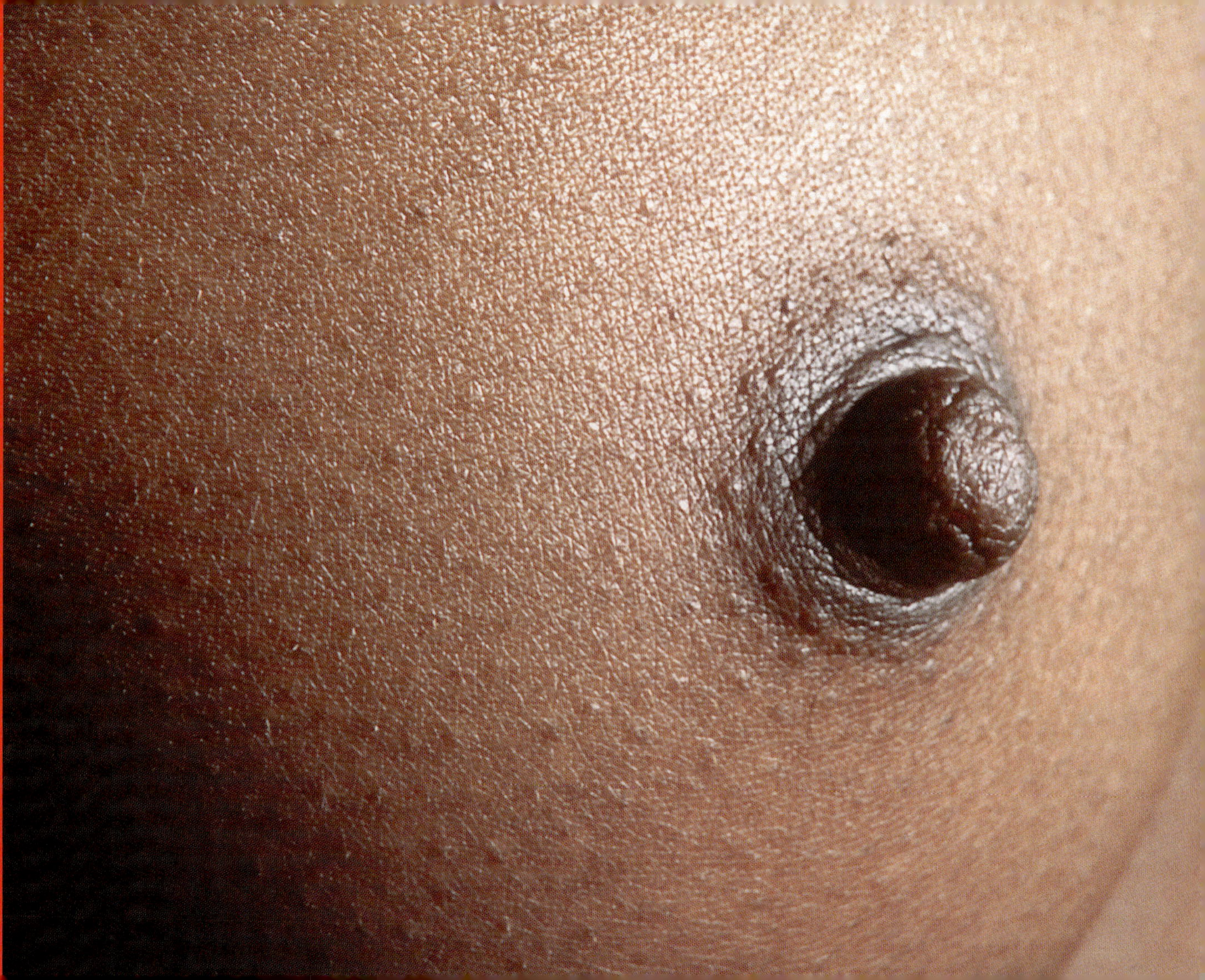

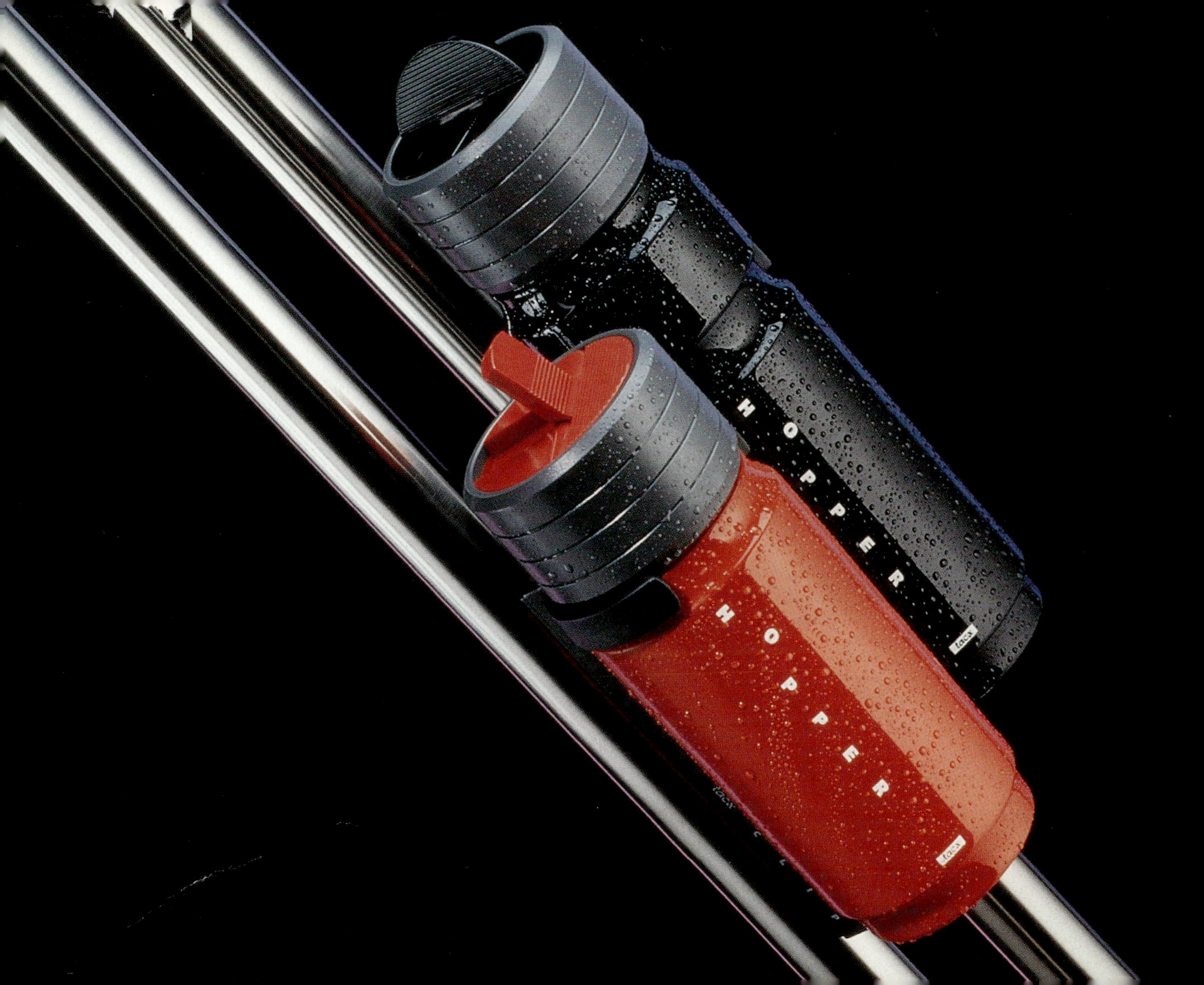
HOPPER
HOPPER
tacx
tacx

NL

Men sen bereik je bu
Buitenreclame! Bel 020 - 562 0 562

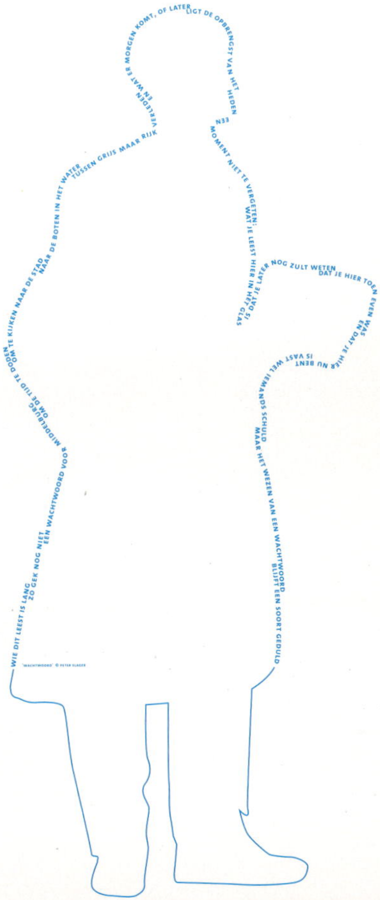
WIE DIT LEEST IS LANG
ZO GEK NOG NIET
EEN WACHTWOORD VOOR
MIDDELBURG
OM DE TIJD TE DODEN
OM TE KIJKEN NAAR DE STAD
NAAR DE BOTEN IN HET WATER
TUSSEN GRIJS MAAR RIJK
VERLEDEN
EN WAT ER MORGEN KOMT, OF LATER
LIGT DE OPBRENGST VAN HET
HEDEN
EEN
MOMENT NIET TE VERGETEN:
WAT JE LEEST HIER IN HET GLAS
IS DAT JE LATER NOG ZULT WETEN
DAT JE HIER TOEN EVEN WAS
EN DAT JE HIER NU BENT
IS VAST
WEL IEMANDS SCHULD
MAAR HET WEZEN VAN EEN WACHTWOORD
BLIJFT EEN SOORT GEDULD
'WACHTWOORD' © PETER SLAGER

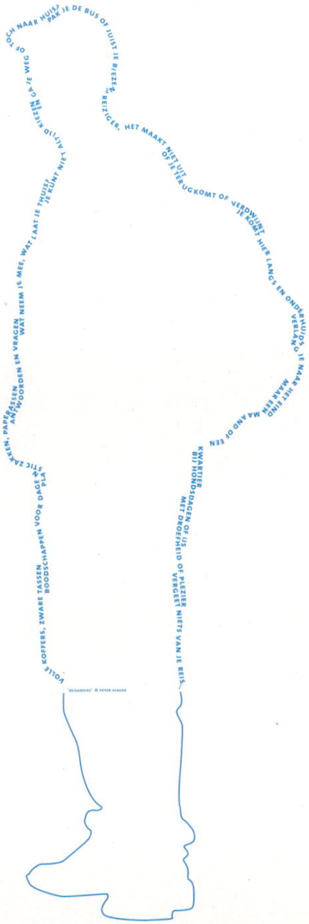
VOLLE KOFFERS, ZWARE TASSEN
BOODSCHAPPEN VOOR DAGEN
PLASTIC ZAKKEN, PAPERASSEN
ANTWOORDEN EN VRAGEN
WAT NEEM JE MEE, WAT LAAT JE THUIS?
JE KUNT NIET ALTIJD KIEZEN
GA JE WEG OF TOCH NAAR HUIS?
PAK JE DE BUS OF JUIST JE BIEZEN?
REIZIGER, HET MAAKT NIET UIT
OF JE TERUGKOMT OF VERDWIJNT
JE KOMT HIER LANGS EN ONDERHUIDS
VERLANG JE NAAR HET EIND
MAAR EEN MAAND OF EEN
KWARTIER
BIJ HONDSDAGEN OF IJS
MET DROEFHEID OF PLEZIER
VERGEET NIETS VAN JE REIS...
'REISADVIES' © PETER SLAGER

NL

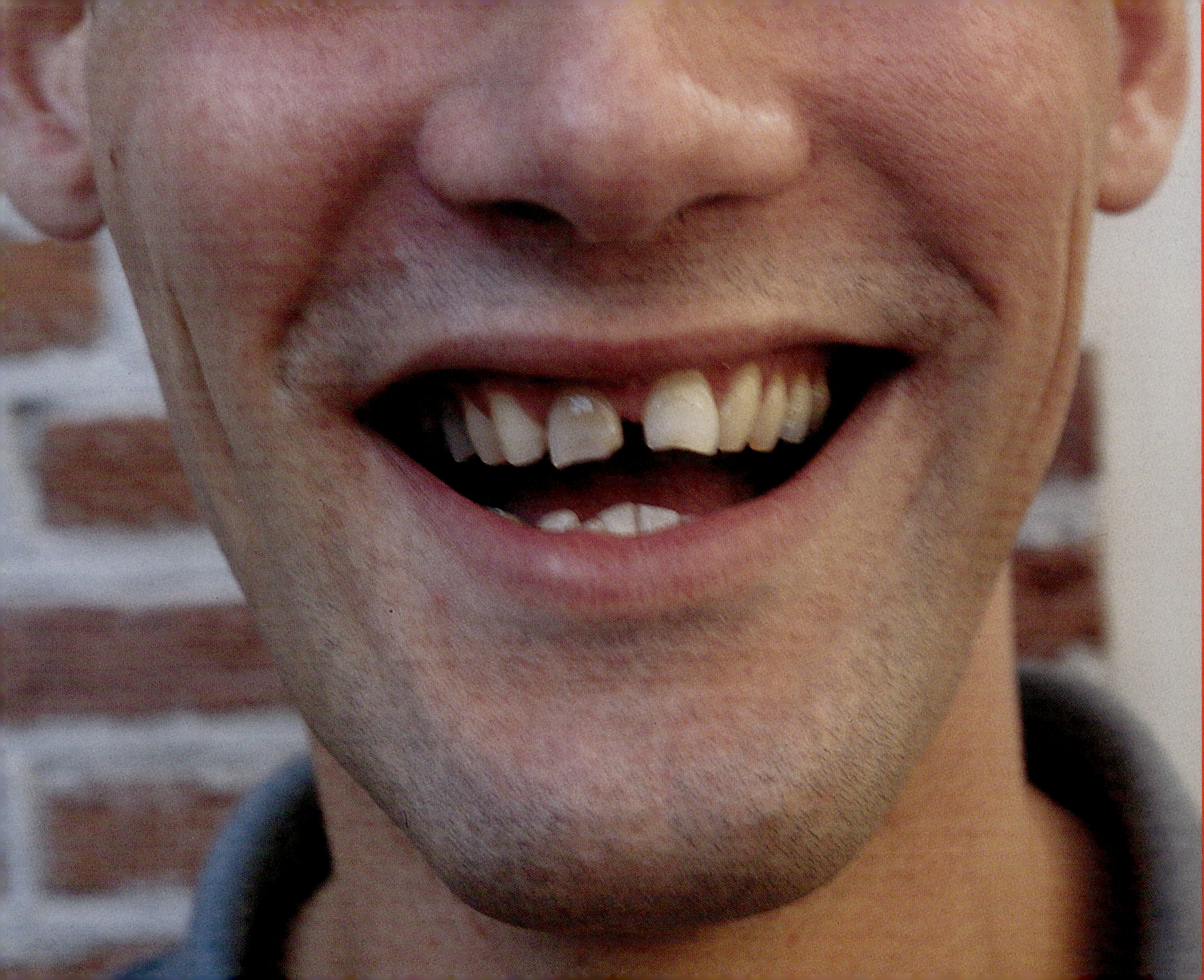

Ritme van de Dag
bij het Dak

Kamperland
Domburg
N 255
400 m
12345/678
ANWB

LAREN
HUIZEN
LAGE VUURSCHE 7,1
BAARN 9,1
AMERSFOORT 18,7
UTRECHT 15,6
LOOSDRECHT 5,-
's GRAVENLAND 4,1
AMSTERDAM 28
VREELAND 11,4
A.N.W.B. TOERISTENBOND

Autosattlerei
A.Hegner
DAMEN
Coiffeur
HERREN

IJ tram
elke 5 min
Kattenburgerstraat
VAN NELLE KOFFIE
KRACHTIGE KOFFIE

contact

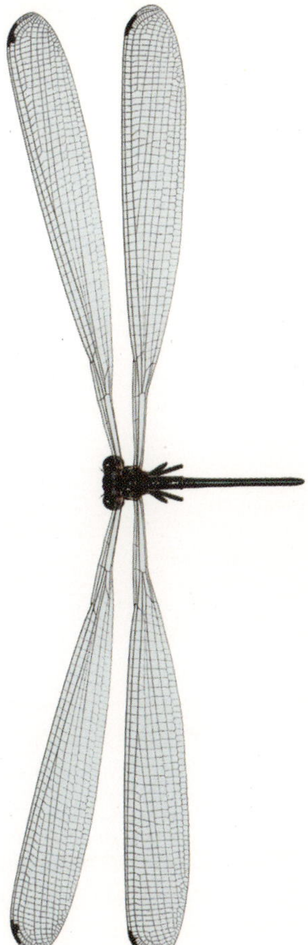

42A

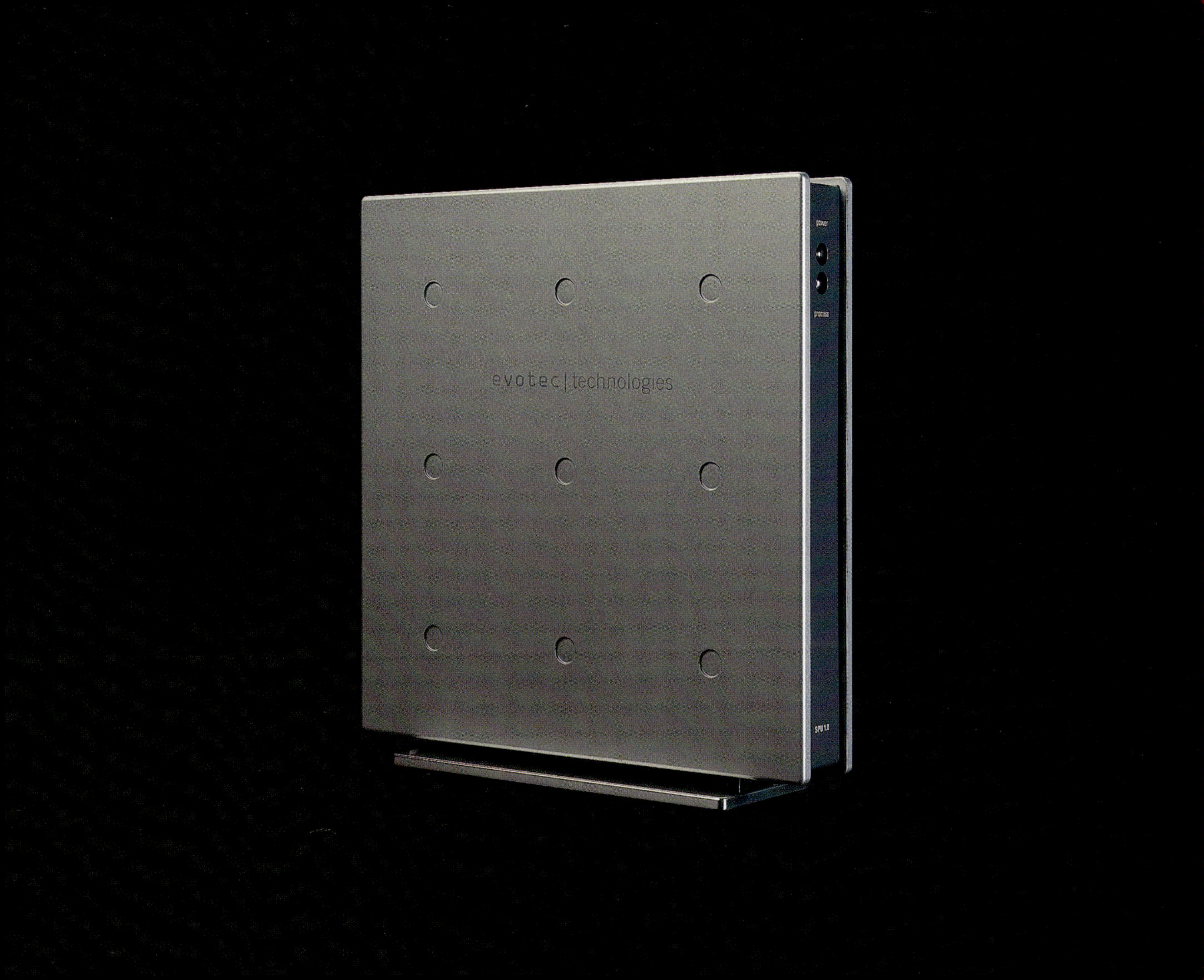
evotec|technologies
power
process
SPU 1.0

hi! r u ok? don't
b l8 2nite! c u @
home b4 9. bbq
sounds gr8! ;-)
xxx

Options

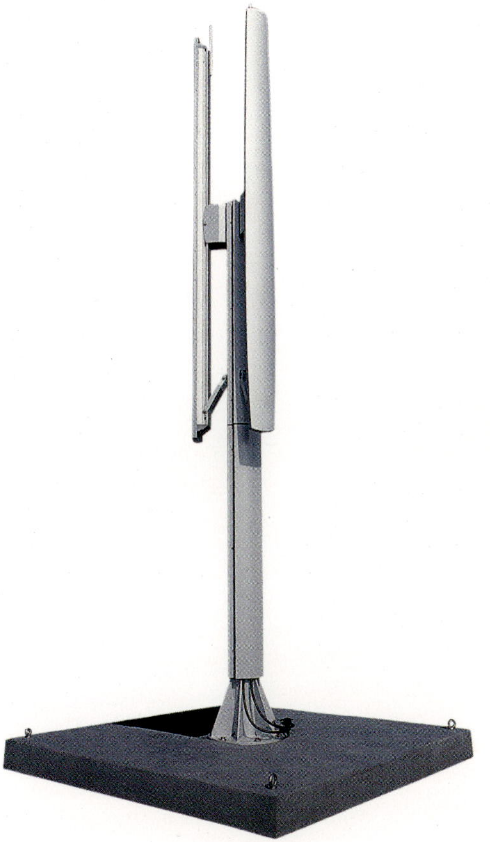

edding 345
copylighter
edding 345
edding 345
highlighter
edding 345
highlighter
edding 345
highlighter
edding 345
highlighter

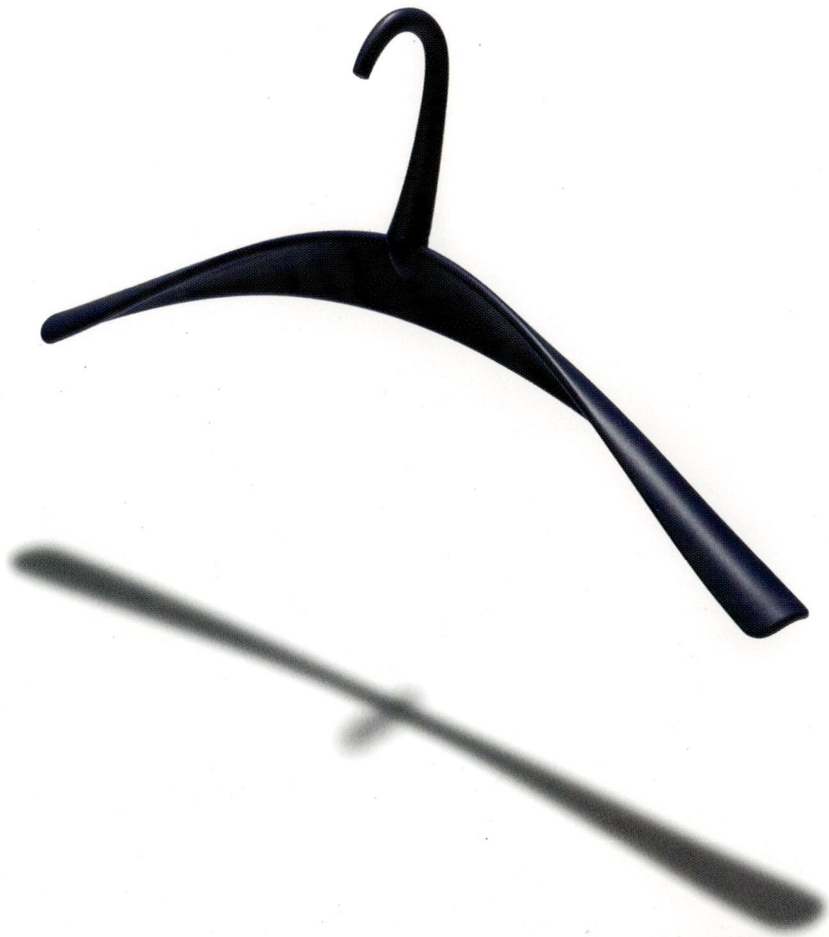

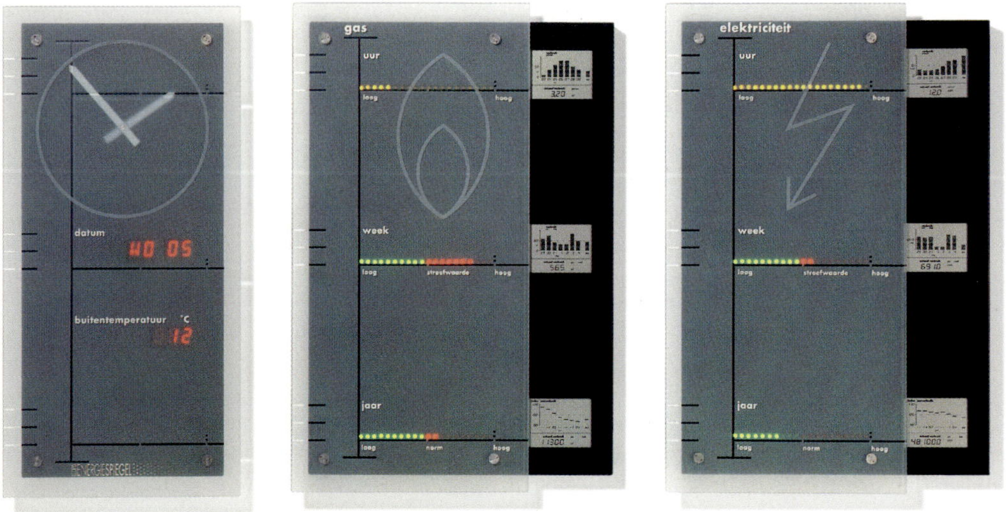
datum
40 05
buitentemperatuur °C
12
ENERGIESPIEGEL
gas
uur
laag
hoog
week
streefwaarde
jaar
norm
elektriciteit

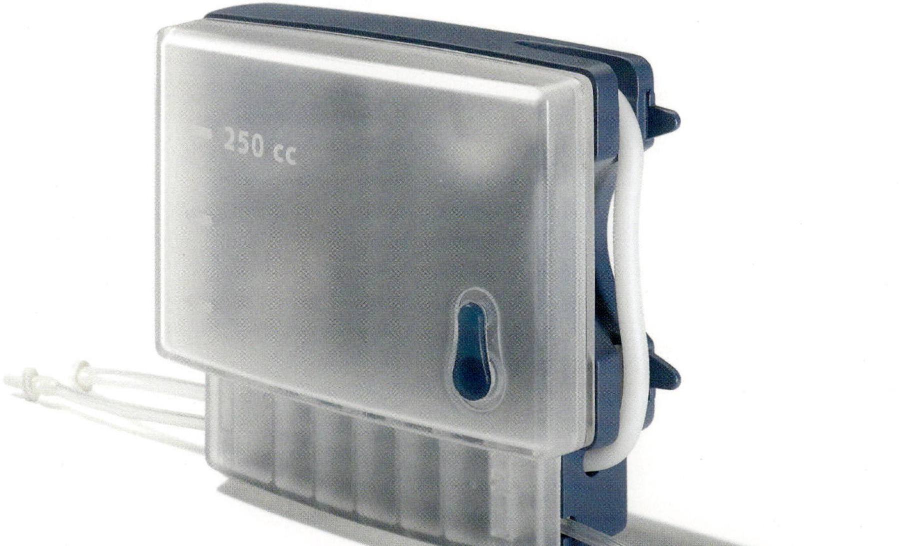
250 cc

electrotherapy device

Fig. 87.

play

Tacx

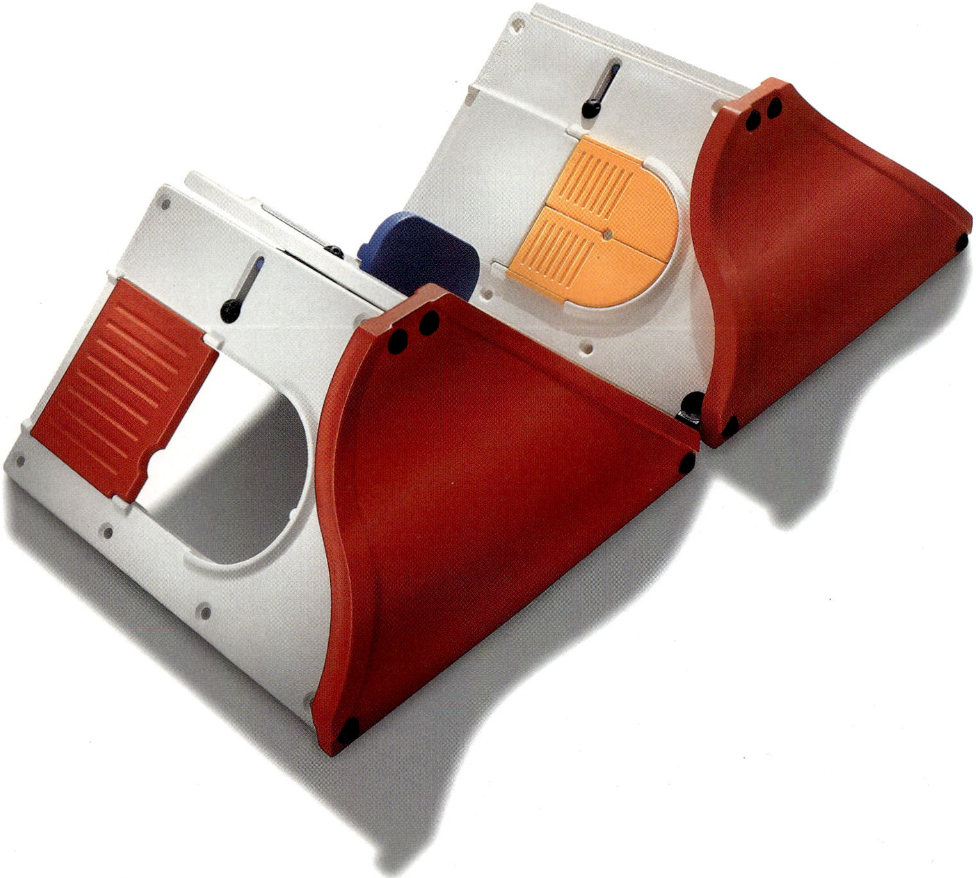

BAAS
EIGEN
Baas
EIGEN
BAAS IN EIGEN

Tacx
CYCLEFORCE
Tacx

rest

seca

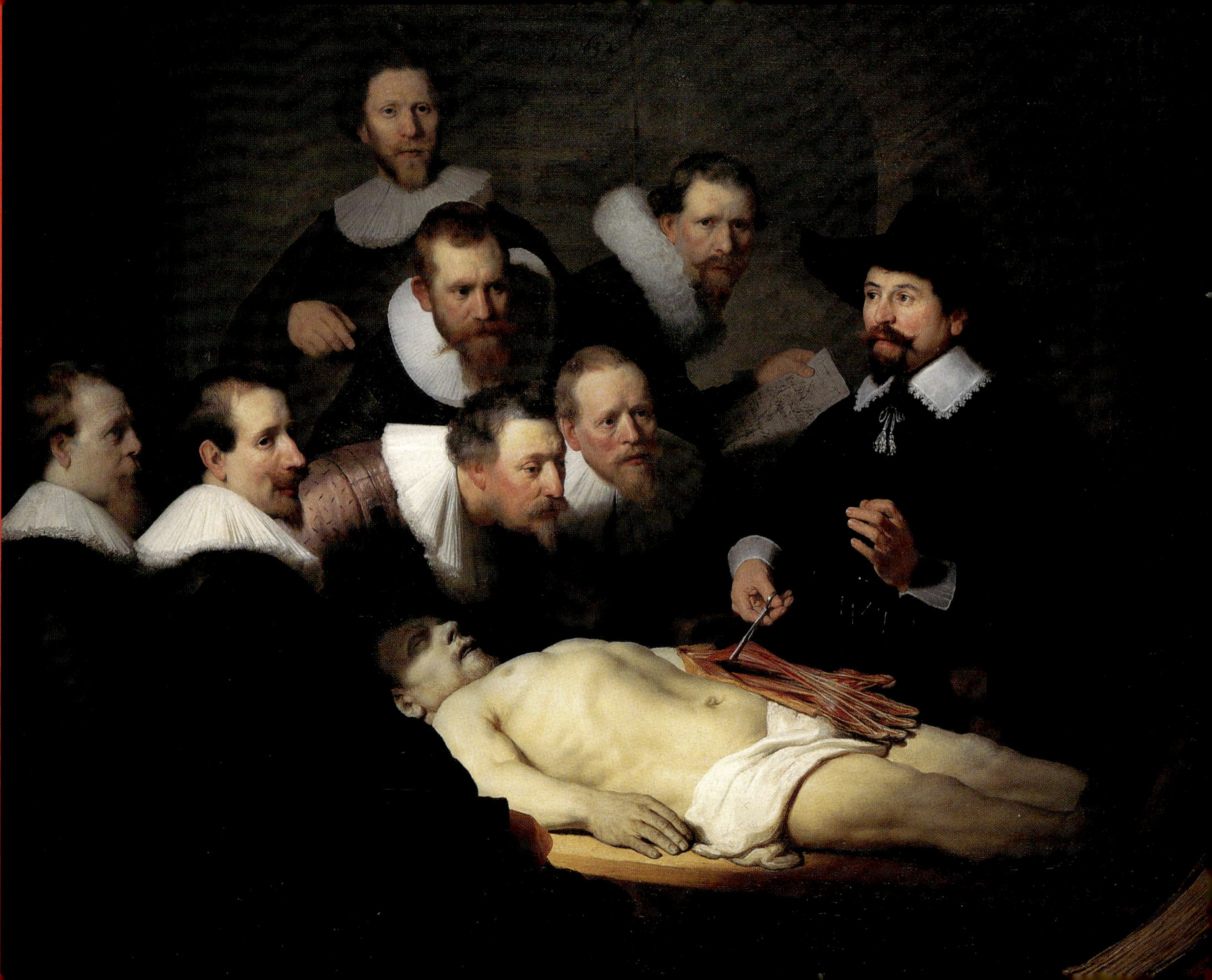

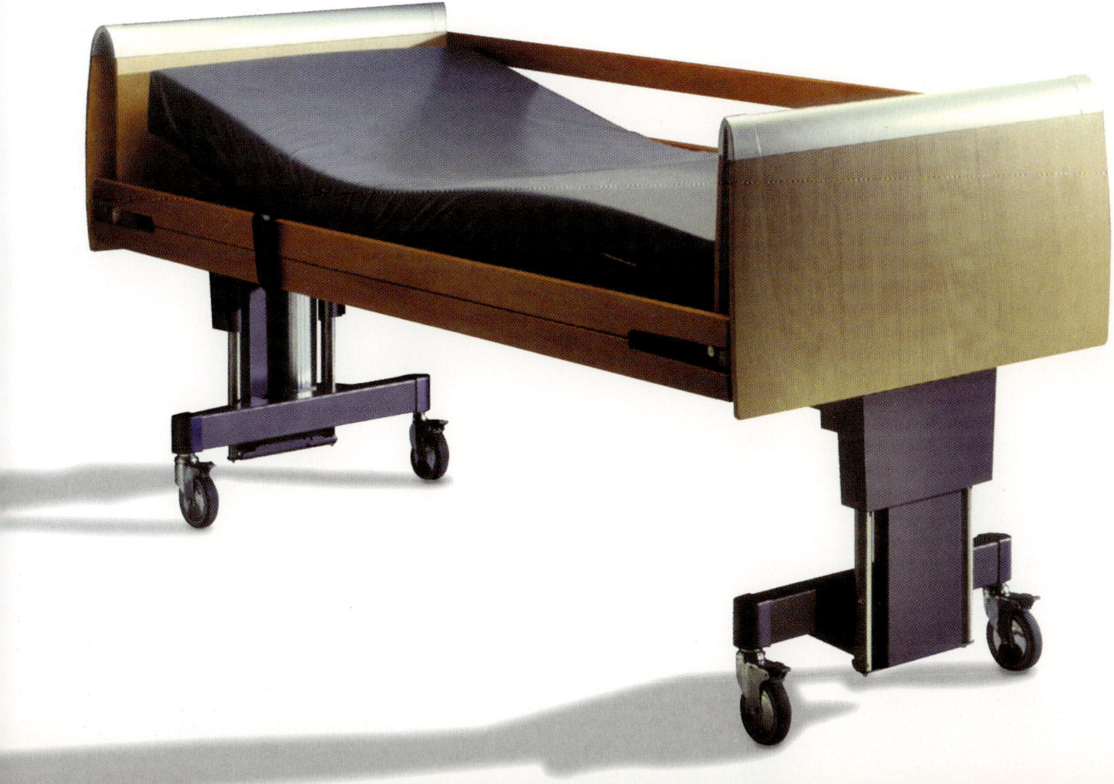

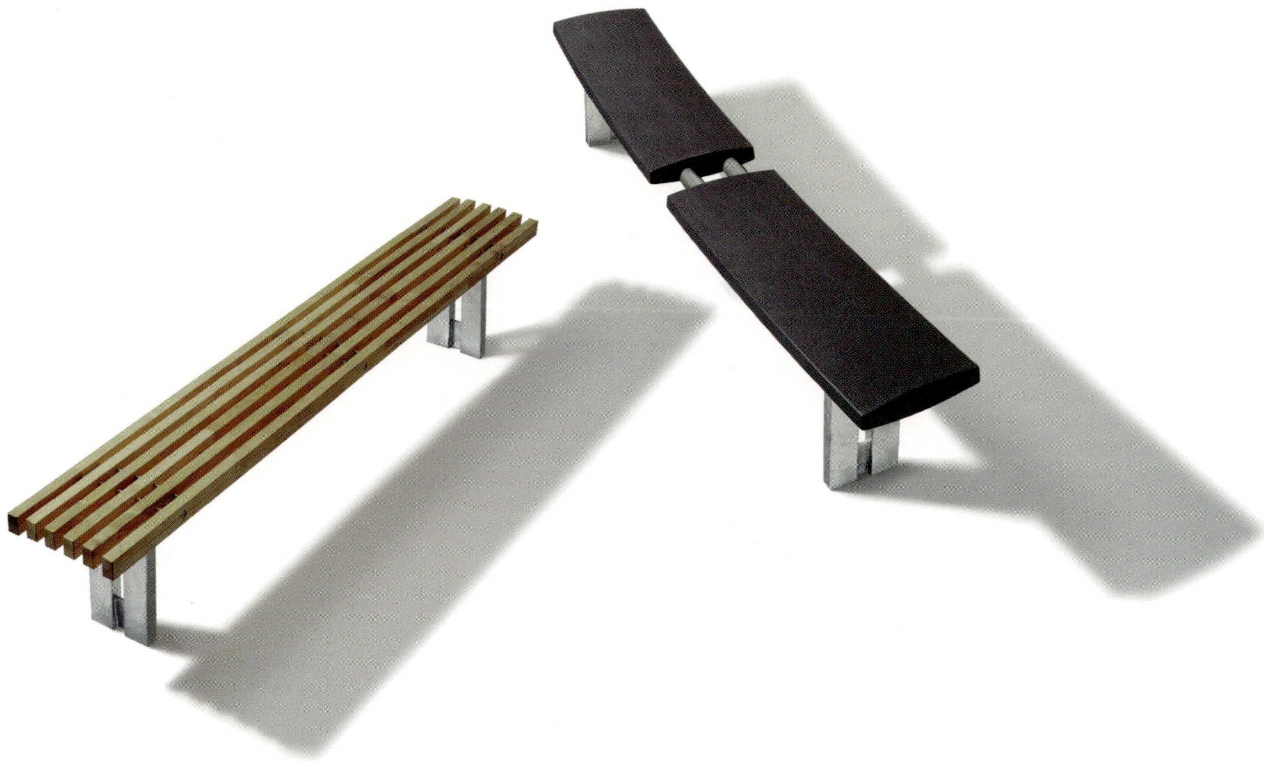

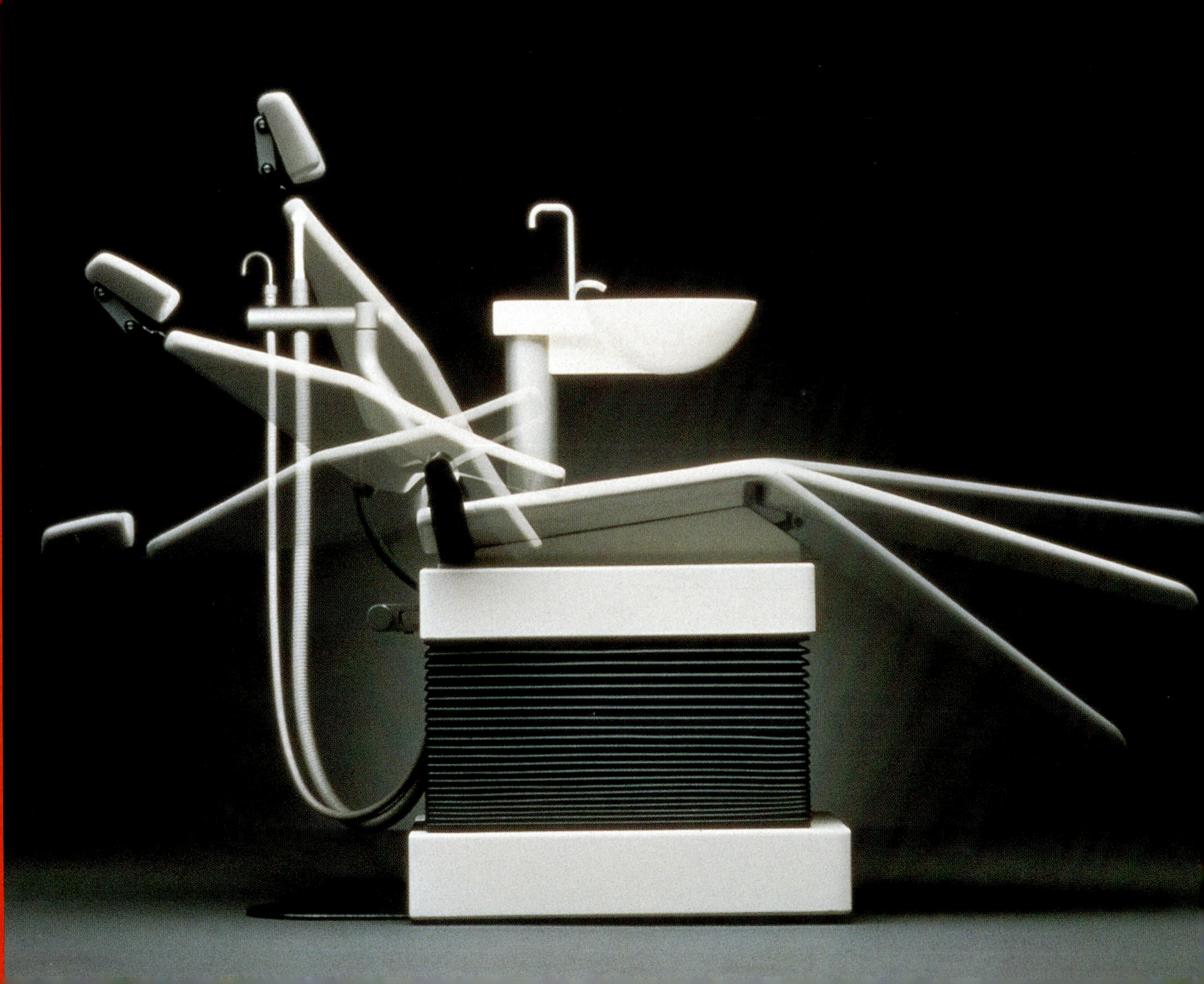

light

Leiden 4
Witte Poort 4

WILA

Friendship

Maastricht

むじんくん
3F
カラオケ
ヨドバシカメラ
モバイル館
ヨドバシカメラ時計総合館
手打ち
らーめん
新高揚
買取
地下1階
Canon
EF LENS

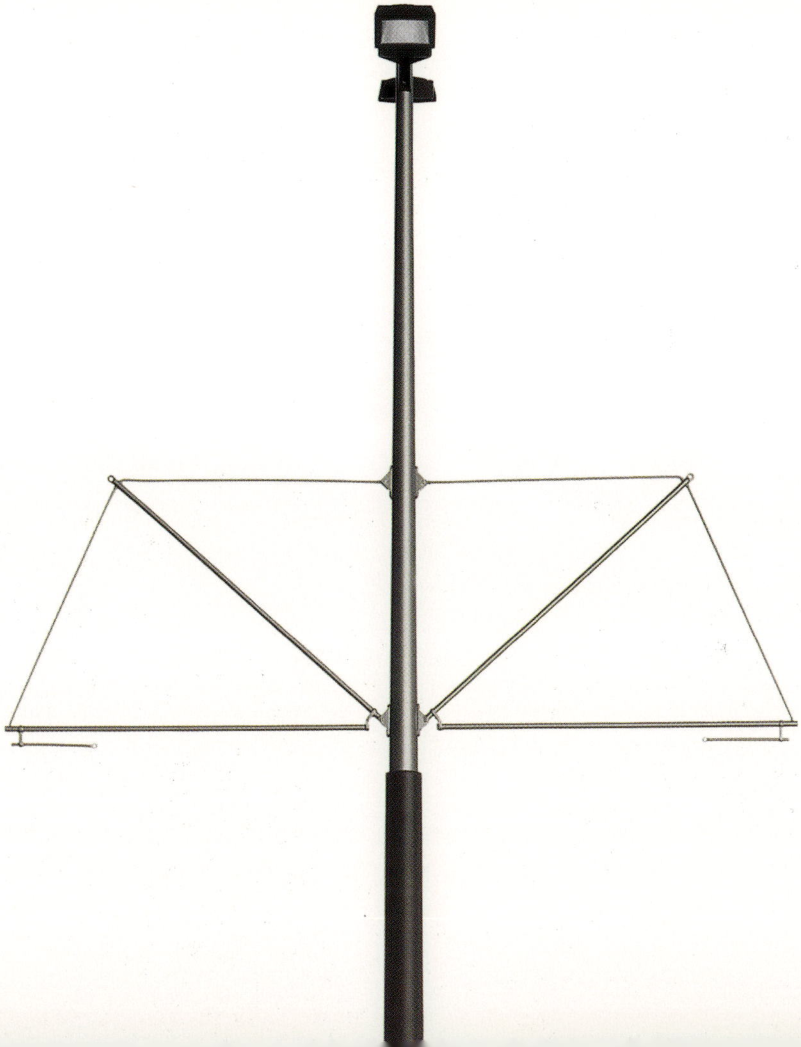

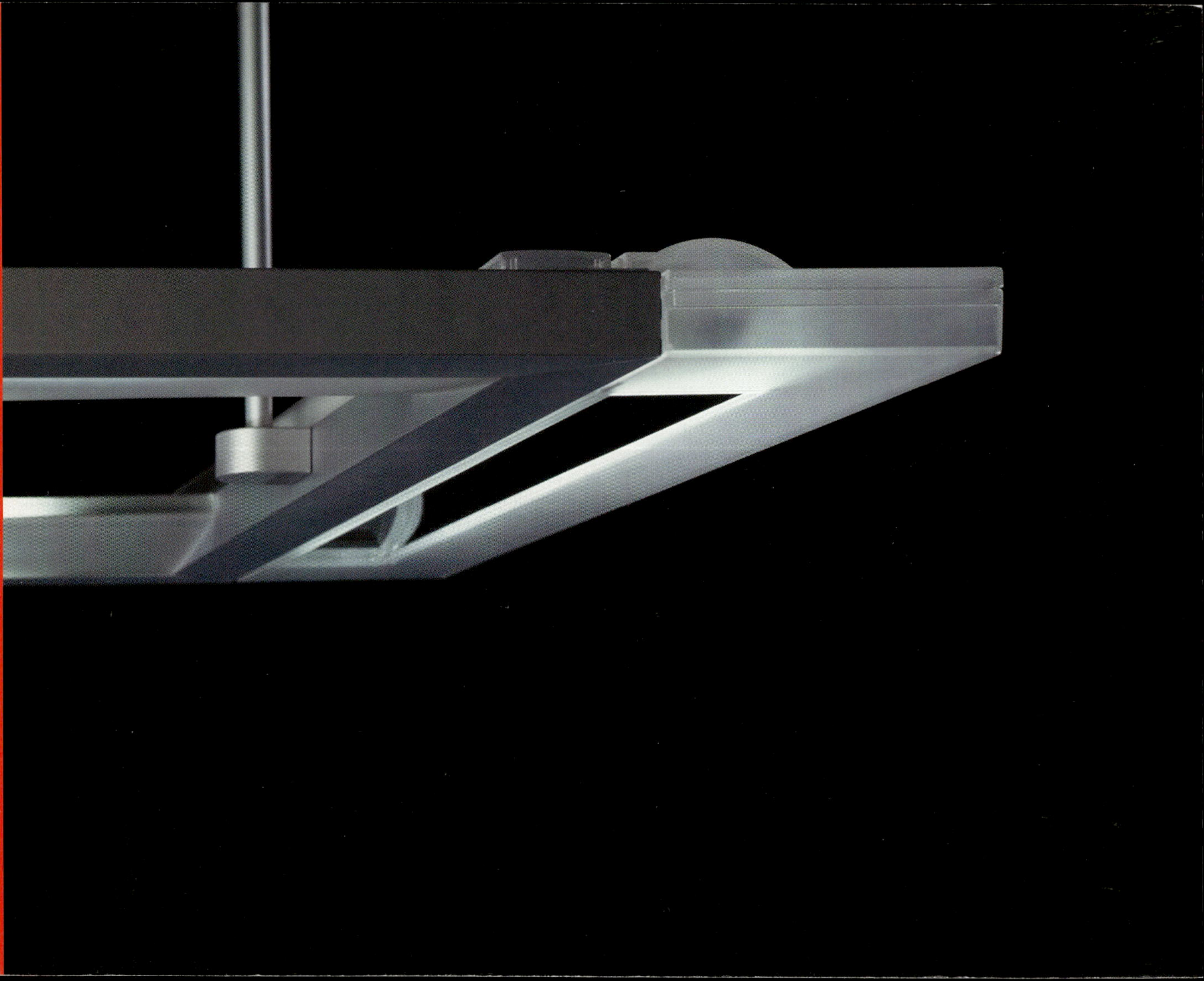

npk people

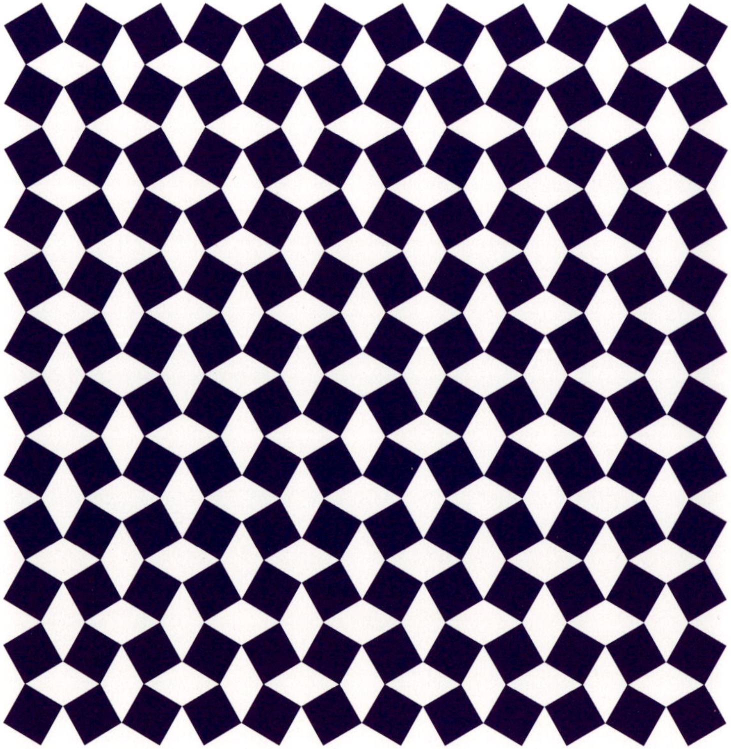

'all for one, one for all' 1998

49
KODAK TX 6043

‘pass it on’ 1991

CLEAN

13 **Stony coral, Indonesia**

14 **Kimberly-Clark, 'Aqua' soap and tissue dispenser,** 2000
GIO award
They belong to the standard equipment in many hotels and hospitals. The oval-shaped plastic housing enables the dispenser to be cleaned with a single wipe. All superfluous details have been banished.

15 **NKL Contactlenzen, Lens mailer,** 1997
Gouden Noot Award
Designprijs Rotterdam, nomination
Design Distinction Award, ID-magazine New York
This aesthetic, light replacement for the well-known, small, round boxes has been made for the purpose of dispatching hard lenses from the factory to the optician as inexpensively and as safely as possible. There is a removable clip on the packing with information for the optician. Removal of the clip makes it a fine lens holder for home use. The leak-proof, plastic packing can be sent in an ordinary envelope. The lens is fixed in the closure cap. A minimum amount of liquid is all that is required to guarantee safe transport.

18 **FEI Company, Electron microscope system XL800 for the Semicon Industry,** 2000
iF award, Hannover
npk industrial design developed the housing for the XL800: a diagnostic dual ion beam microscope, used in clean room laboratories in order to test the quality of semiconductor chips. Machine safety and serviceability were key issues. Compliant with all clean room specifications, the housing is also adaptable to FEI's future technology road map.

19 **Kimberly Clark, 'Ripple' tissue dispenser,** 2000
GIO award
The Ripple product range consists of a number of dispensers for paper hand towels, toilet paper, and soap. The units have been designed specifically for the use in hotels and restaurants. The engineering is of such high quality that it has been possible to reduce drastically the amount of components in the product range.

22 **ASM Europe/ASM International, Vertical Furnace A412 for the Semicon Industry,** 1997
npk industrial design works closely with ASM Europe and ASM International (NL/USA), two high-tech companies which focus on developing, manufacturing and marketing state-of-the-art equipment for the semiconductor industry. The machine is used as a furnace to heat 300 mm wafers and to evaporate gases. npk industrial design delivered its wide range of services, extending from component lay-out, design and pre-construction to the graphic design of the machine documentation.

23 **Chrompack International/Varian, Study of gas chromatograph, 'CP Future',** 1997
Gas chromotagraphy is an essential measuring technique used to create a cleaner and green world. The CP Future is the gas chromatograph of the future: faster, easier to use, more reliable and environment-friendly. The technology ensures a 90% reduction in gas and energy used during processing.

26 **Widget, Towel hook,** 2002
Since 1995, Widget has been engaged in making innovative, colourful consumer products aimed at a young target group. npk industrial design played an important role in the field of strategic design and product development, and designed a large number of Widget bathroom accessories such as bathmat, toilet-roll holder, soap container and towel hook.

27 **Vital Scientific, 'XL Selectra' clinical chemistry analyser,** 2001
Vital Scientific's Selectra XL tests blood or urine samples for various substances, enzymes, medicine medication levels, or drugs. The system is used by hospitals and small laboratories and also by institutions such as prisons. The processing speed is an astonishing 480 tests per hour.

28 **SEM of human red and white blood cells and platelets**

29 **Georgia, Atlanta, USA, highway and skyline at night**

31 **ANWB, four types of finger signs,** 1997
GIO award, special award for Excellence 'Respect'
European Design Annual, Design Excellence
npk industrial design successfully redesigned the four types of the ANWB finger signposts. Although the prevalent colour of the ANWB signing is blue, the bicycle signs are red. The new signs are clearer, more consistent, and more legible. The typography was designed in close co-operation with Gerard Unger. He redesigned the ANWB typeface. Extensive research has shown that this regular graphic design enhances the legibility and comprehensibility of the information.

32 **Bicycle rack in Leiden, the Netherlands,** 2002

34 **Van Riesen, Safety buckle for a child's car-seat,** 2002
Children and safety is a hot issue in the mobile society of today.
No compromises are tolerated regarding safety aspects and functional demands in the design of this safety buckle.

35 **High-Speed Rail (Hogesnelheidslijn),** 1999-2001
The Netherlands' greatest civil engineering project! The high-speed train is rapidly conquering Europe. The 100-km-long track between Antwerp and Amsterdam will make the Netherlands part of this network. npk industrial design has been asked by the HSL-South project team to be actively involved in the development and design of this track.

37 **Jan van Eyck, Christ in Glory, Polyptych of the mystic lamb (central panel), St. Bavon Cathedral, Ghent,** 1432

38 **Agenzia Romana per la Preparazione del Giubileo, Orientation and information system for the city of Rome,** 1999
In close co-operation with Gerard Unger, npk industrial design designed the system for the 'Guibileo' in the year 2000. The highly legible typeface strikes a perfect balance between history and modern design principles.

39 **ASM Lithography, Air cargo container for waferstepper,** 1999
Not only has npk industrial design been involved in the development of several semiconductor machines for ASML, it has also developed several peripheral products such as an air cargo transport container for the Twinscan lithography machine. During transport the container protects and ensures the safe delivery of this machine, sensitive down to nano-levels, by controlling humidity, temperature, pressure, and airflow levels. The air cargo container is also important in communicating the ASML corporate identity worldwide.

42 **Port of Rotterdam/Streetlife, Information lectern,** 1996
The largest port in the world is characterised by an enormous surface area of water and quays that afford numerous interesting views.
npk industrial design designed information lecterns located near the main viewpoints, providing information on the inspiring Rotterdam skyline and the many different vessels and objects on the river. The lecterns are renowned for their clear graphic design and numerous abstract illustrations.

43 **Deutsche Bahn (German Railways), Study for modular platform system,** 2000
'Research on how universal station roofing can be designed for smaller stations at the lowest possible cost price.' This was, in a nutshell, the challenge npk industrial design received from the Deutsche Bahn.
npk industrial design opted for a modular construction system.
Both the supporting sections and the modular roofs are prefabricated and assembled at the location itself while the station is closed for only a very short time.

44 **Train in Tongi, Bangladesh,** 2000
Departure of hundreds of thousands of Muslims after the 'Biswa Istema' Islamic assembly in Bangladesh.

46 **City of Amsterdam, Guide lines for the blind and partially sighted,** 2001
Commissioned by the Welfare Department of Amsterdam and in conjunction with the Stichting Gehandicapten Overleg Amsterdam (Foundation for Consultation with the Disabled), npk industrial design designed a standard system of caution and guide lines for the blind and partially sighted. It is a lucid system that is well suited to the public space.

WORK

47 **Verolme Shipyard by Aart Klein, Rotterdam,** 1962

49 **Tati, film still from the movie Playtime, 'Monsieur Hulot searches for his business partner',** 1967

50 **Kembo, 'Plato' office furniture system,** 1995
GIO award
Plato is an extensive and simple office furniture programme that satisfies the highest expectations in terms of working conditions, ergonomics and the work environment.

51 **dyes Haworth, 'Point' table system,** 1998
The combination of flexibility and ergonomics makes this a high-grade and affordable system. It allows for series production and assembly within a very short time.

53 **Daphne's kimono from Japan**

54 **Janome, 'Quilter's Companian Series' Sewing machine,** 2002
Sewing machine magnate Janome (Japanese for 'eye of the snake') asked npk industrial design to design a family of sewing machines for the European and American markets. The working surface is slightly curved and thus accentuates the flow of the material. The identity of smooth simplicity is combined with the perfection of Japanese fine mechanics.

55 **(UMS)Pastoe, 'Perception' cupboard system,** 1993
iF Best of Category, Hannover
Design Selection Award, ID-magazine New York
Red Dot award, Essen, Germany
Designprijs Rotterdam, nomination
Aluminium Award
Perception is made up of slim aluminium shelves and multiplex side panels that are fitted with hinged doors, drawers, and rear panels in various finishes. It is simple to alter the appearance of the Perception system thanks to the variety of surface finishes and colour combinations.

56 **Extruded Aluminium Sections, Aluminium Pavilion, Swiss National Exhibition,** 1939

58 **Steelcase, 'Let's B' office chair,** 2001
iF award, Hannover
GIO award
These office chairs were designed for the world's largest producer of office chairs. The challenge was to develop an extremely low-priced series of chairs with high-quality aesthetics and numerous ergonomic functions. After a strategic preliminary study, the accompanying visitor's chairs and stools were also realised. The dual chair-back is very distinctive due to the application of two different soft materials for upper and lower lumbar support.

59 **ABN AMRO, Signage system for headquarters,** 1998
iF award, Hannover
The signage for the new head office of the ABN AMRO bank was designed by npk industrial design. Graphic design and industrial design were conceived and carried out as integral parts. The aluminium system, which is based on the building's architecture rather than on corporate identity, is characterised by clarity and simplicity.

62 **Randstad, Office accessories,** 1992
iF award, Hannover
GIO award, Shortlist Minerva Award, London
International Design Yearbook
Thanks to their class and style, these functional durable accessories have received great appreciation throughout Randstad, an employment agency for temporary workers. The various accessories are autonomous and idiosyncratic in design, yet at the same time they function as members of a product family.

63 **S+B, 'Part and Smart System' laboratory furniture and fume cupboard,** 1992
iF award, Hannover
GIO award
Part System is laboratory furniture that latches onto the enormous degree of automation in laboratories. The system provides opportunities for connecting all sorts of media and electrical supplies, cupboards, shelves, and PCs. The Smart fume cupboard guarantees optimal safety for the user by means of optimised internal aerodynamics.

64 **Windtunnel test with a model of the F27 Friendship prototype,** 1955
From: Salto Mortale, Fokker in bedrijf 1991-1996.

EAT

65 **npk industrial design millennium dinner at the Pieterskerk, Leiden,** 2000

67 **Avery Berkel, Meat slicer,** 1996
For Avery Berkel, one of the leading cutting machine manufacturers in the world, npk industrial design designed a machine which strikes a perfect balance between safety, hygiene, and user comfort. Prototypes have been tested by butchers in various user trials in four countries.

68 **Meat mosaic: 'Marble floor' by Wim Delvoye,** 1999

69 **Loladina with chocolate mouth,** 2002

70 **Q for Quality,** 2000
This chocolate plaquette is a delicious promotional gift, inspired by npk industrial design's special signposting design for the City of Rome for the Giubileo 2000. npk industrial design's graphic design section designed the tablet and the packaging.

71 **Port of Rotterdam/Streetlife, Picnic table,** 1997
This picnic table for 12 people is ideal for open-air use and can also be used as an outdoor conference table for companies.

73 **Notice in the 'Roos' bar in Amsterdam,** 2002

74 **DECS/Sara Lee, 'Contenta' coffee dispenser,** 1999
The coffee machine at work is what the water well used to be in bygone days: the place where people meet to drink and talk. This coffee machine is fitted with a compact plastic housing containing a system that provides a supply of approximately 230 cups of fresh coffee.

75 **Solvay, Durable pill case,** 1994
This pill case is attractive yet discrete. The pills are enclosed in a durable and adjustable holder with a strong identity.

78 **Tacx, 'Hopper' water bottle,** 1992
iF award, Hannover
Designprijs Rotterdam (honorable mention in industrial design)
International Design Yearbook
The design is an attempt to reverse the unwelcome but increasing trend towards throwaway products in the Western World.

NL

81 **TPG Post, Twin letterbox,** 2002
The new TPG Post eye-catcher is excellently suited to the allure of the company: reliable and customer-friendly. The new twin letterbox has 50% more capacity and is burglar-proof. The communication has also become more lucid. An electronic display panel between the two openings indicates when the next collection will take place. The functional design consists of fewer components and has been constructed to accommodate technological evolution.

83 **East Asian swallowtail 'Papilio xuthus Linnaeus,** 1767

84 **Streetlife, Butterfly bridge,** 1998
npk industrial design has developed an industrial bridge system that can be used everywhere thanks to its strength and elegant butterfly-shaped support sections. Two curved intersecting tubes, reconnecting to one another at the extremities, form a stable support for the modular anti-slip deck.

85 **City of Middelburg, Bus station,** 2002
All the street furniture and public elements around the railway station in Middelburg have been designed by npk industrial design. These elements - the roofed-over travellers' waiting area, the bicycle sheds, pergola, fences, lighting masts, and two lifts - form a visually simple and coherent system.

86 **Poem by Peter Slager, singer and lyricist from the band Bløf,** 2002
Graphic design: npk industrial design

87 **Perch**

88 **City of Leiden, Dynamic bus station,** 1998
The dynamic bus station has a clear and compact concept, stimulating fast transfers. Its design complements the architecture and surroundings of the existing railway station. The main structure consists of a 65-metre-long, gently-curving tube held in place at various angles by pressure tubes and tie rods. Special 'flipdot' displays and monitors in the central passenger unit provide information on departures and arrivals.

CONTACT

89 **Streetlife, Bicycle racks,** 1997

90 **Bas with gappy teeth 'fietsenrek' (bicycle rack),** 2002

91 **Close-up crocus, Corporate identity of Floriade,** 2002

92 **Stichting Floriade 2002/Streetlife, Temporary Signage System,** 2002
The latest edition of the world horticultural exhibition has been held in the Haarlemmermeer. Around 2,5 million visitors attended the greatest public event in the Netherlands in 2002. In conjunction with garden and landscape architect Niek Roozen, npk industrial design designed the signposting and the information system for the entire park. The signposting is characterised by an open and transparent design that is excellently suited to the temporary nature of this kind of manifestation.

93 **ANWB, Redesign of national signage system,** 1998-2002
The Netherlands is 'on the move'. After the successful redesign of the ANWB finger post system, npk industrial design concentrated on the total Dutch road signage system. Restructuring the information, using open arrows for directional signs, redesigning the typography (by Gerard Unger) and a new set of pictograms, make road signs in the Netherlands a clear design, more balanced and easier to read.

94 **Old Dutch signpost**

95 **Tram stop in Aarau, Switzerland,** 1954

96 **City of Amsterdam, Vertical elements of IJ-tram,** 1999-2002
In 2002 the Amsterdam IJ-tram will link Amsterdam Central Station with IJ-burg. The Amsterdam Quality Assurance Committee has asked npk industrial design to take care of the identity and design of this 15 kilometre line. The elements for the IJ-tram comprise of a variable tram shelter system, fencing, information systems, electricity and lighting masts. The latter have become the standard new lighting system for throughroads in the whole of Amsterdam.

97 **Cables in Rozenburg, the Netherlands,** 2002

99 **Hawaii, Oahu, North Shore, huge wave ('pipeline')**

100 **Grundig, 'Yachtboy 500' world receiver,** 1993
iF award, Hannover
Red Dot award, Essen, Germany
International Design Yearbook
Special Design Award Chicago
This product combines high quality with the smallest possible housing. The sound quality is of such great importance that the size of the loudspeaker was determined at an early stage in the design.

101 **Laarhoven Design, 'LaPlus' foldable display,** 2001
iF award, Hannover
Stylish presentation at every conceivable location: no problem with LaPlus. The design of this inventive product is modest yet exceptional. It folds in like an umbrella and can easily be carried in a special transport casing. An easily replaced print - paper, synthetic material or textile - is spanned on a cross-structure of thin fiberglass tubes. Several displays together can form an entire display wall.

104 **Evotec Technologies, Signal Processing Unit (SPU),** 2002
The SPU is a key component for biomedical microplate readers. npk industrial design developed a housing for the SPU that improves analytical processes in the pharmaceutical industry. With this design, Evotec technologies established a new direction for professional devices in the field of biotechnology: a high-quality product that combines technical performance with state-of-the-art technology.

105 **Siemens, Home Delivery Box,** 2001
The Siemens Home Delivery Box avoids the need to stay home waiting for articles ordered via Internet. The Box consists of three compartments and is suitable for goods such as groceries, post packages and medicine. The refrigerating and freezing compartment provides space for fresh and frozen groceries. The Box can hold a substantial quantity of shopping and also serves as a repository.

107 **'Hi! Are you ok? Don't be late tonight! See you at home before nine. Barbecue sounds great! kisses'**

108 **Vodafone, 'Flex Solution' antenna system,** 2000
iF award, Hannover
Although mobile telephone antennae are considered by some to be an eyesore in the built-up environment, Vodafone has evolved beyond this status. npk industrial design has developed an antenna system characterised by an elegant and unobstrusive design, enabling the antennae to blend in with the surrounding architecture. The main benefits for the client are the fact that the system is both cost-effective and timesaving. The system and its accessories are prefabricated and easily installed, with the added advantage of reducing any nuisance for residents to a minimum.

109 **Edding, Text markers,** 2002
The worldwide sales figures of the new 345 text marker exceed all expectations. The slender elliptical design gives the marker a modern and elegant look. The product comprises only three components and has a recognisable identity. Besides the usual five neon colours, a special grey tint is also on offer. This grey tint is unique because it now enables marked texts to be photocopied.

110 **Purple and yellow tulips in the Hazepolder,** the Netherlands

111 **Maria Blaisse, Rubber hat for Issey Miyake,** 1990

112 **MK Tresmer, 'Swan' coat hanger,** 1999
npk industrial design created an elegant plastic coat hanger for this Finnish manufacturing company. The distinctive twisting shape makes this product surprisingly strong despite a minimal use of materials.

113 **Ecofys, Energy consumption panel,** 1996
International Design Yearbook
Designprijs Rotterdam, nomination
A good example of a product with a convenient user interface design. The Energiespiegel (energy panel) allows users of a building to acquire information on the consumption of gas or electricity, or on current temperature at a single glance.
The easy-to-read LED bar-graph behind the sandblasted glass plate gives a rough impression of energy consumption, while the three LCD displays in the steel housing providemore detailed information.

114 **Californian Windmills,** 1990

115 **Marjan's blue eye,** 2002

116 **Dorc International, Liquid cartridge of the 'Associate' optical surgical station,** 2001
In eye operations, nothing is left to chance. The Associate is an operating theatre unit to which all kinds of optical surgical instruments can be connected. The innovative cartridge controls the streams of liquid by means of a fluid-regulating system of ducts and valves, a vacuum sensor, a filter, and peristalsis unit.

117 **Uniphy, 'Phyaction 787' electrotherapy device,** 1988
iF Best of Category, Hannover
The Phyaction is a user-friendly electrotherapy device for physiotherapists. The user-interface is multi-lingual and fully menu-controlled. npk industrial design supplied the basic layout, the design and the engineering of the housing and front panels.

118 **Medico-mechanical equipment of Doctor Gustav Zander,** 1865
This equipment can be found in the permanent collection at the Museum Boerhaave, Leiden, the Netherlands.

PLAY

121 **A woman tobogganing down the Dorf Dimson bobsleigh run at St. Moritz, Switzerland**

122 **Hamax, 'Sno-shark' snow sledge,** 1999
Designprijs Rotterdam, nomination
The Sno-shark made by Hamax, the Norwegian world market leader in sledges, is an innovative one-child sledge with excellent gliding and steering properties. The design of the Sno-shark appeals to a child's imagination, with its animated character and car-like presence. The high gloss accentuates the round forms and gives an impression of speed.

123 **Tacx, 'Tao' bottle cage for bicycles,** 2002
Many top teams participated in the 2002 Tour de France with the ultralight Tao bottle cage fixed to the frame. The Tao is a functional adornment to the bicycle. The cage is punched from an aluminium plate, formed, anodised, and partially sprayed with a soft synthetic material.

124 **'De piste van Diavolo', from catalogue 'De fiets', an exhibition of Museum Boijmans Van Beuningen in Rotterdam,** 1977

125 **Eupholus beattle from Biak**

126 **Hamax, 'Sno-cross' snow sledge,** 2002
The latest sledge developed by npk industrial design is called Sno-cross. The design of the Sno-cross looks like a race motorbike, on purpose because it is made for tough kids. After several tests, Sno-cross turned out to be the fastest sledge of all! The sledge is easy to transport to the slope because it is fully foldable and fits into every ski-box.

127 **Curver, 'Cosy Casa' Wendy house,** 1997
The Good Toy Guide Awards - Silver Award
Produkte des Jahres 1998 - Fachverband Kunststoff-Konsumwaren im GKV, Frankfurt
Curver was looking for new products for outdoor use that could be sold by existing department stores and garden centres. In close conjunction with play psychologists npk industrial design developed a modular system of large outdoor play equipment. The 50 colourful blow-moulded components can be easily combined in various ways not just by parents, but also by children. This solid, safe play equipment stimulates the imagination and creativity.

128 **Beach cabins at Katwijk, the Netherlands,** 1982

129 **Dolle Mina 'Baas in eigen buik',** 1970
In October 1970, Dolle Mina organised a demonstration through the inner city of Amsterdam in favour of the contraceptive pill. The contraceptive pill was introduced in the early 1960s, but was only covered by the Dutch National Health Service in 1971.

130 **Organon, NuvaTime®,** 2002
The NuvaTime acts as a reminder tool for bi-monthly use of a new contraceptive by using universal imagery to indicate appropriate action. Multi-continental user research has demonstrated that women appreciate the NuvaTime as a convenient, appealing, and useful memory aid that enhances their sense of comfort, security, and control.

131 **Tacx, 'CycleForce Excel' cycle trainer,** 1997
iF Best of Category, Hannover
Designprijs Rotterdam, nomination
International Design Yearbook
Now everyone has the opportunity to work seriously on his or her fitness at home with the Tacx CycleForce, a new generation of cycle trainers. The handsome and solidly built stand is fitted with an innovative folding mechanism that makes it easy to store. The stand is linked via an electromagnetic system to an Excel computer that records both performance and heartbeat data.

133 **Dutch football fans celebrating**

134 **ANWB, The national pictogram set for The Netherlands,** 2002
The 'Stadium' pictogram is part of the new national pictogram set for The Netherlands, designed by the graphic design team of npk industrial design.

135 **Sea lions from Patagonia, Argentina**

137 **Streetlife, 'Wait&See' urban bench,** 1999
The 'Wait&See' system consists of stainless steel supported with SMC or Robinia wooden seats.

138 **Man on woodpile, Bruynzeel Deurenfabriek by Piet Zwart,** 1931

139 **Pheasant's nest on Texel, the Netherlands**

140 **Dremefa, 'Multibob' child seat,** 2002
Comfort, ease of use, and above all safety: these were the most important starting points in designing the Multibob. An ingenious mechanism ensures that when the height of the head support is determined, the back support is automatically set to the proper ergonomic proportion. The arm supports function as lateral 'airbags' because they fold in towards the child in case of a collision.

141 **Seca, 'Chica' baby weighing scales,** 2000
Red Dot award, Essen, Germany
npk industrial design designed fold-up baby weighing scales that are not only used in the Western world but also in developing countries. The scales can be suspended on a special carrier.

144 **Kembo, 'Thema' conference chair,** 1994
GIO award
Thema is a beautiful, timeless conference chair that is the product of close collaboration with the manufacturer, Kembo. The combination of steel, leather and aluminium lends the chair a certain elegance. In addition, a great deal of attention was paid to ergonomic aspects in the design. The chair has a comfortable but active seat.

145 **Etesmi/JWKoch, 'Plano' hospital bed and bedside cabinet,** 1992
iF Top 10, Hannover
iF Best of Category, Hannover (for bedside cabinet)
Plano was designed with a view to increasing both the patient's independence and the effectiveness of the nursing staff. What is more, it does not require a great deal of maintenance and it is easy to clean. The same well thought-out styling and material choice have also been applied to the matching bedside cabinet.

147 **Retaining wall of the Deltawerken in Zeeland, the Netherlands**

148 **Streetlife, 'Wait&See' public seating system,** 1999
A seat system specifically designed for public transport areas.

149 **Kembo, 'Chiaro' office chair,** 1995
GIO award
iF award, Hannover
Design Selection Award, ID-magazine New York (honorable mention in furniture products)
The Chiaro is a practical and multifunctional chair designed for the project market. The austere design has a dynamic quality thanks to the distinctive slim line and the almost casual curves. This solid and comfortable chair is made of aluminium and wood and comes in different colours and models.

150 **A roof section of Casa Milá (La Pedrera) by Gaudí, Barcelona, Spain**
Casa Milà is an apartment building situated in Barcelona. It was built between 1906 and 1910 for the Mila family and is the last example of Gaudi's architecture.

151 **Rembrandt van Rijn, 'De anatomische les van Dr. Nicolaes Tulp',** 1632

152 **Etesmi/JWKoch, 'Ampio' nursing-home bed,** 1999
The Ampio was designed primarily with nursing homes in mind. Head and foot-ends consist of curved wooden boards connected by an elliptical aluminium profile used as grip rail. Each end has runners for the wooden side rails, integrated in between the curves. The bed looks inviting and natural in its lower positions, whilst the higher positions provide the nursing staff with a comfortable working height.

153 **Streetlife, 'Wait&See',** 1998

154 **Whippet Bench by Radi Designers, Paris,** 1998
The shape of the bench was generated by extruding a whippet's profile. The use of colour reinforces the dynamic effect of the freeze-framed image of the dog.

156 **Hamax, 'Kiss' bicycle seat,** 1996
GIO award
iF award, Hannover
Norwegian Design Award
Design Selection Award, ID-magazine New York (honorable mention in consumer products)
npk industrial design has a great deal of experience in the design and development of child seats. This bicycle seat is made for children weighing up to 22 kilograms (6 years) and is easy to install and to adjust. The position of the support guarantees the child a perfect sitting posture and gives more space to the feet of the cyclist.

157 **Holland Dental, 'Luna 1' dentist's chair,** 1991
iF award, Hannover
The Luna 1 concept consists of a dentist's chair with a basin and a drive for instruments. The system meets the stringent criteria concerning ergonomics, hygiene, and environmental issues. The construction is extremely compact and very user-friendly. All functions are operated by two foot controls. The seat parts are completely seamless, comfortable, strong and easy to disinfect.

LIGHT

159 **City workers protecting their eyes while viewing a solar eclipse from their office roof.**

161 **ANWB, Illuminated finger signs,** 1997

163 **City of Leiden/Streetlife, 'The Bowler' lamppost,** 2002
The elegant 'The Bowler' lighting masts can be deployed at special places, such as Leiden Central Square.

164 **René Magritte, 'Golconde', Oil on canvas, The Menil Collection, Houston, Texas, USA,** 1953

165 **The Milky Way**

166 **WILA Leuchten, 'E-Connect' downlights,** 1996
Kho Liang Ie Design Prize, Wolfram Peters and Jos Oberdorf
iF award, Hannover
Red Dot award, Essen, Germany
GIO award
European Engineering Award
The E-connect system is a modular system of built-in lighting fittings for various diameters/types of PL bulbs. Thanks to the sophisticated system design, a wide range of fittings can be created with a minimum of components. The parts consist of aluminium, magnesium, and plastic.

167 **WILA Leuchten, 'CircLet' lighting system,** 1997
Design Selection Award, ID-magazine New York (honorable mention in furniture)
GIO award
npk industrial design developed Circlet, a sober and compact series of fittings, for the special purpose of installing them in the ceiling or suspending them on a lower level. A great deal of time and effort went into designing the lighting technique, involving the application of new reflectors and anti-dazzle grills, which guarantees a maximum light emission without eye irritation. The lighting series has three variations: built-in, mounted, and suspended.

168 **The ZK-BXB, the F27 Friendship of the National Air Corporation of New Zealand,** 1961
From: Salto Mortale, Fokker in bedrijf 1991-1996.

170 **National Design Museum New York, Solar Pergola System,** 1995
The Solar Pergola System is a clever combination of a nicely designed urban space and a solar unit connected to the mains. The design was developed as a concept for the New York National Design Museum's 'Under the Sun' exhibition, summer 1998.

171 **City of Maastricht/Streetlife, 'Light Frame' information unit,** 1999
The light frame is a new interpretation of the traditional illuminated information unit found mounted on masts and façades in and around towns and cities. The design re-addresses a number of product characteristics to achieve an efficient and contemporary design.

172 **City of Tokyo, Japan,** 2002

174 **City of Amsterdam, Tram and lighting masts,** 1999
Along the new Amsterdam IJ-tramline, npk industrial design was responsible for the identity and design of all vertical elements, from fences to overhead cable masts. These striking masts are fitted with a system of internal counterweights, which maintain the required tension in the overhead cables. The masts are also lighting masts, so that the tramline is also floodlit. Amsterdam municipal council was so pleased with the design that it awarded npk industrial design the contract of designing a new street lighting system in the same style. This has now become the new standard lighting system for thoroughfares in the whole of Amsterdam.

175 **WILA Leuchten, 'Sonic' office lighting system,** 1999
Design Plus Award
A well-lit desk makes a world of difference. This flat, suspended fitting is equipped with TL5 fluorescent bulbs, combining direct and indirect lighting. Thanks to the square orientation of the light lines, the fitting provides anti-dazzle light for twin work places in office buildings. The open structure and the use of semi-translucent plastic enhance the transparent design. The highly sophisticated and elegant perforation creates a beautiful pattern.

176 **Studio Drive-In, Culver City, USA,** 1993

npkPEOPLE

179 **npk industrial design employees, 'Made in Holland',** 2002

180 **Stack of wooden shoes, the Netherlands**

182 **npk industrial design employees, 'All for one, one for all',** 1998

183 **npk industrial design employees, 'Twenty-six hours a day',** 1993

184 **Inner court and entrance to npk industrial design office (listed building from the 17th century), Leiden,** 2002

185 **Nautilus from the Indian Ocean**

186 **npk industrial design employees, 'Pass it on',** 1991

In 1991 nodigde het Stedelijk Museum in Amsterdam npk industrial design uit om een keuze te tonen uit hun toen al indrukwekkende oeuvre. Met enige aarzeling accepteerde npk industrial design de uitnodiging, want in hun ogen was een openbare verantwoording juist op die plaats niet zonder risico. De tentoonstelling 'Function Forms Style' en de daarbij behorende catalogus zouden immers ijkpunten zijn waaraan de verdere ontwikkeling van het bureau zou kunnen worden afgemeten. Een museale verantwoording zou hun werk ook plaatsen in een wel heel specifieke context van kunst en uniciteit die ze eigenlijk steeds hadden weten te ontlopen. Ze wilden en willen geen solistische kunstenaars zijn, maar een gelijkwaardige partner van de industriële opdrachtgever.

Er is inmiddels heel veel gebeurd en nu komt npk industrial design zelf uit met het boek Vision & Precision om opnieuw in het openbaar verantwoording af te leggen met een keuze uit het werk van de laatste tien jaar. Wat zou er na die toen door publiek en professie zo bewonderde tussenstand nog aan ruimte zijn voor verdere groei en verdieping van de mentaliteit en werkwijze van npk industrial design?

npk industrial design werd in 1978 opgericht en is uitgegroeid van een tweemansbureau van de oprichters Peter Krouwel en Wolfram Peters tot een bureau met vijf partners en zestig hooggekwalificeerde medewerkers. Nadat in 1997 Bruno Ninaber van Eyben als partner het bureau verliet, zijn Jan Witte en Jos Oberdorf als nieuwe partners tot het Nederlandse bureau toegetreden en Detlef Rhein tot het Duitse bureau. Direct vanaf het begin heeft npk industrial design zich gericht op het ontwerpen van technisch en functioneel hoogwaardige producten voor gebruik door mensen in hun woon- en werkomgeving en in wat meestal de openbare ruimte wordt genoemd, zoals de straat en het openbaar vervoer. Het gaat hier om producten die in grote series op industriële schaal worden geproduceerd of waarvan maar enkele stuks worden vervaardigd. Er zijn ontwerpen bij met de maat van een microchip en heel grote, zoals de tramlijn naar IJburg, de nieuwe stad in het IJsselmeer bij Amsterdam.

Professionaliteit van micro- tot macroschaal

Het is altijd een bewuste keuze van npk industrial design geweest om alles tussen micro- en macroschaal aan te kunnen. Dit is echter niet zo eenvoudig. Kleine producten dwingen de ontwerper tot de precisie en detaillering van een sieradenmaker. Grootschalige ontwerpen vereisen de kracht van het grote gebaar om eenheid en overzicht in de visuele expressie te behouden.
Het werken aan grootschalige producten doet ervaren dat visuele nadruk op de millimeter afbreuk doet aan het overzicht, terwijl kleinschalige producten aantonen dat het grote gebaar tot in de detaillering zichtbaar moet zijn. Deze wisselende schaal is niet alleen bedoeld om de ontwerpers te dwingen regelmatig met een ander oog te kijken. Het gaat ook om het vakmanschap in de breedte en zeker ook om het pure genoegen van de professionele uitdaging.

Productietechnisch zijn er ook grote verschillen, zodat de technische expertise van het bureau dezelfde breedte moet bestrijken. Hierin gaat npk industrial design heel ver.
De modelmakerij is uitgegroeid tot een plek met de allure van een laboratorium, waar producten tot in ieder detail kunnen worden gemaakt en getest. Hier worden met de hand of met geavanceerde computergestuurde machines de prototypen en nulseries vervaardigd waarmee de productie van een ontwerp wordt uitgeprobeerd en voorbereid. Vanuit gespecialiseerde teams wordt de vervaardiging aangestuurd van de matrijzen en andere gereedschappen om de ontwerpen elders te kunnen produceren. Dit betekent dat npk industrial design tot het allerlaatste moment de controle op de kwaliteit van ieder aspect van het product behoudt.

De gedachte achter deze breedte van micro- tot macroschaal en de behoefte om alle technische aspecten in eigen hand te houden, is dat van het werken op verschillende niveaus van complexiteit, functionaliteit, schaal en technologie een inspirerende invloed en interactie uitgaat. Het dwingt tot het steeds opnieuw zoeken naar een vormgeving die in balans is met de maat en de functie van het product. Deze bandbreedte is een opmerkelijke kwaliteit van npk industrial design. Het is zelfs de vraag of er elders een ontwerpbureau gevonden kan worden dat dezelfde ervaring en kennis bezit om deze verscheidenheid zo professioneel en zo consequent aan te kunnen. Regelmatig gaat het om nieuwe producten waarvan weinig of geen voorbeelden bestaan en waarvoor nog veel uitgezocht moet worden. Het zijn altijd complexe producten waarin veel nieuwe kennis is verwerkt en die niet in een achternamiddag bedacht kunnen worden, maar die het resultaat zijn van veel onderzoek en experiment.

De eisen waaraan dergelijke ontwerpen moeten voldoen, of ze nu groot zijn of minuscuul, maakt het noodzakelijk systematisch te werk te gaan om het spoor niet bijster te raken. Hiervoor maakt npk industrial design gebruik van een inmiddels beproefde methode waarin een ontwerpprobleem een reeks van stadia doorloopt die begint met onderzoek en analyse van het probleem en die geleidelijk steeds concreter wordt met schetsen en modellen. Het hart van dit proces is het stadium waarin alle voorwaarden in een optimale samenhang in elkaar worden geschoven tot wat een ontwerp in zijn meest elementaire en beschrijvende vorm is: een materieel voorbeeld dat industrieel gereproduceerd kan worden. Het gaat echter niet alleen om de technisch-industriële maakbaarheid of de functionaliteit van een ontwerp, hoe belangrijk die ook zijn. De vormgeving is zeker zo belangrijk en mogelijk zelfs het belangrijkste aspect.
De vorm wordt niet alleen als eerste waargenomen, maar is tevens de materiële drager voor alle functionele eigenschappen. Het uiterlijk van een product heeft ook een culturele dimensie waarmee het een eigen plek in de tijd krijgt. Dit lijkt wel een autonoom proces te zijn dat uitstijgt boven de functie en de maakbaarheid van het product. Bovendien lijkt het nauwelijks rationeel te beïnvloeden, want zelfs de meest nuchtere vormgever is uiteindelijk een kind van zijn tijd en daaraan gebonden door zijn beleving van de wereld om hem heen.

De grootste kracht van npk industrial design is dat het in staat is om een heldere integratie te bereiken van de drie belangrijkste aspecten van een product: de industriële maakbaarheid, de functionaliteit en de vormgeving met daarin besloten de culturele dimensie. Dit is een moeizaam en tijdrovend proces van kritisch zoeken en proberen en afwijzen en opnieuw beginnen en doorgaan tot alles op zijn plek valt. Opdrachtgevers hebben begrijpelijke haast en zullen vaak verbaasd en mogelijk zelfs ongeduldig zijn over dit eigenwijze doorploegen tot de laatste millimeter, die soms wel erg ver weg lijkt te liggen. Het feit dat npk industrial design deze weerbarstige gedrevenheid al 25 jaar met succes weet vol te houden tegenover haar opdrachtgevers, toont dat de resultaten overtuigend zijn.
Er zijn ook internationaal maar weinig bureaus waar zo klinisch en systematisch, maar tegelijk zo intens en gedreven wordt ontworpen en waar de zelfkritiek zo groot is en zo lang wordt volgehouden. De wens van perfectie en de behoefte om alles uit een ontwerp te halen, maakt dat npk industrial design nog weleens op eigen initiatief en risico buiten de specificaties van een opdracht wil stappen om te zien of alle mogelijkheden wel uitgezocht zijn, al is het maar voor de eigen zekerheid.
Soms zijn de grenzen al bereikt en is er geen nieuwe stap meer te maken. Het komt echter ook regelmatig voor dat een opdrachtgever verrast wordt met een nieuw idee waarmee zijn product een extra dimensie in kwaliteit kan krijgen.

Ontwikkelingen in diepte en breedte
Sinds de tentoonstelling in het Stedelijk Museum is npk industrial design in diepte en breedte gegroeid. Een logische ontwikkeling is dat er andere vakgebieden zijn bijgekomen, zoals grafische vormgeving. Veel producten dragen grafische elementen in de vorm van merknamen en merkbeelden, aanduidingen voor de bediening en er zijn handleidingen nodig. Bij veel producten verloopt de interactie van bediening en controle tussen gebruiker en product met wat wel wordt aangeduid als 'user interface'. Voor al deze aspecten is een afdeling grafische vormgeving opgericht, die inmiddels een eigen dynamiek heeft gekregen die verdergaat dan alleen de ondersteuning van de productontwerpers. Voor specialistische projecten wordt echter intensief samengewerkt met professionals van buiten, zoals met typograaf Gerard Unger voor de nieuwe landelijke bewegwijzering voor de ANWB.

Uit de persoonlijke belangstelling voor producten in de openbare ruimte bij een van de partners is een nieuwe productenlijn gegroeid met banken, straatverlichting en wegwijzersystemen. Inmiddels is hierin zoveel kennis en ervaring opgebouwd, dat npk industrial design in staat is de hele uitrusting van tramlijnen, compleet met halteplaatsen en verlichtingssystemen, in eigen hand te houden. Voor de nieuwe Hogesnelheidslijn (HSL) ontwikkelt npk industrial design behalve de uitrusting ook het bijzondere tracé, dat in hun handen een modulair en industrieel te produceren betonnen product wordt. Traditioneel zou hiervoor een civiel ingenieursbureau zijn gevraagd.

npk industrial design heeft internationaal de vleugels uitgeslagen. Sinds enkele jaren is in Hamburg een vestiging actief om zowel logistiek als cultureel dichter bij de Duitse opdrachtgevers te staan die de weg naar het bureau reeds hadden gevonden. Een vestiging in Boston, in de Verenigde Staten is in voorbereiding.

Naast een verantwoording van een succesvol bureau toont dit boek ook de emotionele interesse die npk industrial design heeft in het medium fotografie, die veel verdergaat dan het documenteren van de eigen ontwerpen. Vanaf het begin heeft het bureau belangstelling voor dit medium, met een eigen professionele fotostudio en het werken met externe fotografen, zoals Gerrit Schreurs. Zijn foto's laten de ontwerpen niet alleen zien door zijn ogen. Ze zijn het resultaat van een diepgaande dialoog met het bureau. Bij eerste beschouwing lijken ze, net zoals de eerste kennismaking met npk industrial design, zakelijk en afstandelijk. Nauwkeurig bekeken, wordt echter zichtbaar hoe door materiaaluitdrukking en een geraffineerd gebruik van schaduw de persoonlijke intenties van de vormgever worden geaccentueerd. Voor het registreren in de eigen studio is samen met Gerrit Schreurs een beeldvorm ontwikkeld waarmee eenheid wordt bereikt in de beeldpresentaties van npk industrial design.

Opzet Vision & Precision
Vision & Precision geeft op een verrassende en directe manier ook inzicht in dieper liggende emotionele overwegingen die een ontwerp bepalen en op een elegante manier worden twee fundamentele problemen opgelost die altijd optreden wanneer we naar afbeeldingen van gebruiksvoorwerpen kijken. Een productafbeelding is in feite een uitgeklede representatie van de werkelijkheid. Allereerst klopt de maat niet en het perspectief in het afgebeelde tweedimensionale beeld verstoort de inschatting van de schaal. De achterkant is niet zichtbaar en ook iedere beweging ontbreekt, terwijl juist veel van onze gebruiksvoorwerpen bewegen om hun functie te kunnen vervullen.

De waarneming van, en dus ons oordeel over veel van deze producten, of het nu een kinderzitje, een stoel of een laboratoriuminstrument is, wordt in belangrijke mate bepaald door tactiele ervaringen en deze ontbreken bij het bekijken van een afbeelding. Aspecten als gewicht en balans, de textuur en de temperatuur van het oppervlak, de weerstand en het geluid van knoppen en hendels, de wisselende reflecties van het licht op het oppervlak zijn niet te ervaren. Wat echter vooral ontbreekt, is de context waarin de afgebeelde producten functioneren, want onze gebruiksvoorwerpen opereren altijd in een omgeving met andere producten.

Er is nog een probleem bij het bekijken van afbeeldingen van producten. Onze seriële gebruiksvoorwerpen hebben in de twintigste eeuw een dubbelrol gekregen van functievervuller én drager van de persoonlijke expressie van de bezitter. Ze spelen een rol in ons leven als nuchtere en trouwe functievervullers en dienen daarmee ons comfort en onze veiligheid bij het verrichten van taken. Daarnaast heeft de vormgeving een opmerkelijke en overheersende rol gekregen in de niet-materiële expressie van ons zelfbeeld, of in het beeld zoals we graag gezien willen worden. Het lijkt er hierbij op alsof de gebruiker een deel van de eigen expressie van zijn zelfbeeld heeft afgestaan aan zijn materiële omgeving. 'Toon mij uw gebruiksvoorwerpen en ik zeg u wie u bent', is geen loze opmerking. Kennelijk kunnen aan hetzelfde product verschillende individueel bepaalde interpretaties gehangen worden. Het gebruiksvoorwerp wordt hiermee een willige drager van een veelheid van betekenissen.
npk industrial design heeft een verrassende en doeltreffende oplossing gevonden om deze beide waarnemingsproblemen

zichtbaar en hanteerbaar te maken voor de gebruiker van dit boek. Iedere afbeelding van een ontwerp in dit boek heeft als een onverbrekelijke pendant een afbeelding naast zich.
De producten worden geïsoleerd getoond, zonder context, maar de pendantafbeelding doet het werk. Deze beelden - meestal foto's - roepen bij de kijker spontaan allerlei persoonlijke interpretaties op waarmee zijn waarneming van de afbeelding van het product wordt ingekleurd. De pendantbeelden maken direct duidelijk dat ieder individu op zijn eigen manier en met zijn kennis en inzicht - en vooral met zijn persoonlijke belevingswereld - de producten om hem heen bekijkt. Duidelijk wordt ook hoe divers onze gebruiksvoorwerpen geïnterpreteerd kunnen worden en hoe weinig de ontwerper daarin kan sturen. Hij kan niet bepalen hoe de gebruikers moeten kijken, want er is gewoon geen eenduidige waarneming mogelijk. Zijn kracht als vormgever moet zijn een zodanige vorm aan te bieden dat die een meerduidige interpretatie ondersteunt zonder aan persoonlijkheid te verliezen. Het gebruik van pendantbeelden heft voor een deel ook het genoemde probleem op van de onvolledige representatie van de werkelijkheid. Dit verschijnsel is nu niet meer zo erg, omdat de spontane associaties een, misschien als virtueel te omschrijven, werkelijkheid oproepen waarin het incomplete beeld een eigen, afgeleide werkelijkheid oproept, die bovendien een emotionele inhoud heeft gekregen.

Het boek maakt dus niet alleen duidelijk wat npk industrial design doet. Op een visuele en aanstekelijke manier wordt de kijker ook bewust gemaakt van de vele en geschakeerde relaties die wij hebben met de honderden verschillende functionele producten in onze dagelijkse omgeving. Vision & Precision is zeker geen plaatjesboek. Bestuderen ervan biedt naast een interessante verantwoording ook een spannende interactie aan tussen beeld en waarnemer, die beseft dat over een maand of een jaar andere associaties kunnen ontstaan, doordat er nieuwe ervaringen zijn opgedaan en meer kennis is verworven.

Reyer Kras

Reyer Kras (1940) studeerde industriële vormgeving en kunstgeschiedenis. Hij gaf les aan de Technische Universiteit Delft (industrieel ontwerpen) en de Rijksuniversiteit Leiden (design geschiedenis). Van 1982 tot 2001 was hij werkzaam als conservator industriële vormgeving in het Stedelijk Museum Amsterdam. In 1996 richtte hij het bureau Aurelius industrial design op voor productontwikkeling en museale projecten. Hij heeft veel gepubliceerd en doceert regelmatig in binnen- en buitenland.

1991 erhielt npk industrial design eine Einladung vom Stedelijk Museum in Amsterdam, eine Auswahl seines, bereits zu diesem Zeitpunkt beeindruckenden, Oeuvres auszustellen. Nach einigem Zögern nahm npk industrial design diese Einladung an. Es war schließlich nicht ohne Risiko, sich in aller Öffentlichkeit und gerade an diesem Ort zu verantworten. Die Ausstellung 'Function Forms Style' und der dazu gehörende Katalog hätten als Eichmarken dienen können, an denen man die weitere Entwicklung des Designbüros gemessen hätte. Darüber hinaus hätte eine museale Verantwortung das Büro in den sehr spezifischen Kontext von Kunst und Unität drängen können, dem es sich bisher lieber entzogen hatte. npk industrial design wollte und will kein solistischer Künstler, sondern gleichwertiger Partner industrieller Auftraggeber sein. Seither hat sich viel ereignet und heute ist es npk industrial design selbst, das das Buch Vision & Precision veröffentlicht, um sich wiederum der Öffentlichkeit mit einer Auswahl aus dem Werk der vergangenen zehn Jahre zu stellen. Welchen Raum würde es nach dem damals von Publikum und Fachkreisen so bewunderten Zwischenstand für weiteres Wachstum und Vertiefung der Mentalität und Arbeitsweise von npk industrial design geben?

npk industrial design wurde 1978 gegründet und ist von dem Kleinbüro der Gründer Peter Krouwel und Wolfram Peters zu einem Büro mit heute fünf Partnern und sechzig hochqualifizierten Mitarbeitern angewachsen. Nachdem 1997 Bruno Ninaber van Eyben als Partner das Büro verließ, traten Jan Witte und Jos Oberdorf als neue Partner dem niederländischen Büro und Detlef Rhein dem deutschen Büro bei. Von Anfang an hat sich npk industrial design auf die Produktentwicklung technisch und funktionell hochwertiger Investitions- und Konsumgüter der Wohn- und Arbeitsumgebung des Menschen, sowie auf den öffentlichen Raum, wie Straßen und öffentliche Verkehrsmittel, spezialisiert. Es handelt sich dabei um Produkte, die in industrieller Serienproduktion hergestellt oder von denen nur einige Exemplare angefertigt werden. Darunter finden sich sowohl Produkte von der Größe eines Mikrochips, als auch Entwürfe sehr großen Umfangs, wie die Straßenbahnlinie nach IJburg, der neuen Stadt im IJsselmeer bei Amsterdam.

Professionalität von Mikro bis Makro

npk industrial design hat sich stets bewusst für die Verwirklichung von Produkten in allen Größenordnungen zwischen Mikro und Makro entschieden. Doch so einfach ist das nicht. Kleine Produkte zwingen den Designer zu der Präzision und der Detailtreue eines Goldschmieds. Entwürfe großen Stils erfordern die Kraft der großen Geste, damit die Einheit und Übersicht in der visuellen Expression gewahrt bleibt. Während der Arbeit an großen Projekten gewinnt man die Einsicht, dass die visuelle Ausreizung bis auf den Millimeter dem Gesamteindruck schadet, während kleine Produkte offenbaren, dass die große Geste bis ins Detail sichtbar sein muss. Die wechselnden Dimensionen bewirken nicht nur, dass der Designer gezwungen ist, immerzu mit anderen Augen zu schauen. Es geht auch um die profunde Kenntnis in der Breite und nicht zuletzt auch um das pure Vergnügen an der professionellen Herausforderung.

Produktionstechnisch sind die Unterschiede ebenso groß, so dass die technische Kompetenz des Büros dieselbe Breite erlangen muss. Die Modellwerkstatt hat sich zu einem Ort mit dem Flair eines Laboratoriums ausgeweitet, in dem Produkte bis in das kleinste Detail gefertigt und getestet werden können. Prototypen und Nullserien, mit denen die Herstellung eines Produktes kontrolliert und vorbereitet wird, werden manuell oder mit fortschrittlicher, computergesteuerter Technik angefertigt. Spezialisten steuern außerdem die externe Anfertigung der zur Produktherstellung benötigten Werkzeuge. Das heißt, dass npk industrial design die Kontrolle über die Qualität eines jeden Aspekts des Produkts behält.

Die Idee hinter diesem breiten Größenspektrum von Mikro bis Makro und das Bedürfnis, sämtliche technische Aspekte selbst zu verantworten, ist die Annahme, dass das Arbeiten auf den unterschiedlichen Ebenen von Komplexität, Funktionalität, Größe und Technologie, Inspiration und Interaktion bewirkt. Der Designer ist angehalten, stets nach einer neuen Formensprache zu suchen, die mit dem Maß und der Funktion des Produkts im Einklang ist. Dieses Spektrum zeichnet die Qualität von npk industrial design aus. Es ist sogar fraglich, ob sich ein anderes Designbüro finden ließe, das über einen vergleichbaren Erfahrungsreichtum mit den entsprechenden Kenntnissen verfügt, um diese Verschiedenheit so professionell und so konsequent bewältigen zu können.

Regelmäßig entwirft das Büro neue Produkte, für die es wenig oder keine Vorlagen gibt, und für die zahlreiche gründliche Untersuchungen durchgeführt werden müssen. Es sind immer komplexe Produkte, in die umfangreiches Wissen eingeflossen ist. Immer sind sie das Ergebnis umfassender Untersuchungen und Experimente. Die Anforderungen, denen solche Produkte gerecht werden müssen, ob sie groß oder winzig klein sind, machen ein systematisches Arbeiten unabdingbar, da man das Ziel sonst leicht aus den Augen verlieren kann. Daher bedient sich npk industrial design einer inzwischen erprobten Methode.
Das Designproblem durchläuft dazu eine Reihe von Stadien, beginnend mit der Problemanalyse, bis es sich mit Hilfe von Skizzen und Modellen allmählich herauskristallisiert.
Das Herz dieses Prozesses bildet das Stadium, in welchem alle Bedingungen in einen möglichst optimalen Zusammenhang ineinander geschoben werden. Jetzt entsteht das, was Design in seiner elementarsten und beschreibenden Form ist: eine Materialvorlage, die industriell reproduziert werden kann.
Es geht allerdings nicht nur um die technisch-industrielle Machbarkeit oder die Funktionalität eines Designs. So wichtig diese Aspekte auch sein mögen. Die Formgebung ist mindestens genauso wichtig, wenn nicht gar am wichtigsten. Schließlich wird die Form nicht nur als erstes wahrgenommen, sie ist überdies materieller Träger aller funktionellen Eigenschaften.
Die Produktgestaltung hat außerdem eine kulturelle Dimension, die dem Produkt einen eigenen Platz in der Zeit zuweist. Das scheint ein autonomer Prozess zu sein, welcher über die Funktion und die Machbarkeit des Produkts hinausgeht. Außerdem ist er offensichtlich kaum rational beeinflussbar. Schließlich ist selbst der nüchternste Designer nur ein Kind seiner Zeit und durch das Erleben der Welt um sich herum an den Zeitgeist gebunden.

Die größte Stärke von npk industrial design liegt in der Fähigkeit, die drei wichtigsten Aspekte eines Produkts klar zu integrieren: industrielle Machbarkeit, hoher Gebrauchswert und ästhetische Formgebung, in welche die kulturelle Dimension eingeschlossen ist. Dies ist ein mühevoller und zeitaufwändiger Prozess. Kritisches Suchen und Versuchen wird von Verwerfen, wieder von vorn Beginnen und Fortführen so lange abgewechselt, bis alles im Einklang ist. Auftraggeber haben verständliche Eile und werden über all dies eigensinnige Durchpflügen bis zum letzten Millimeter, der häufig arg weit weg zu sein scheint, oft überrascht und möglicherweise sogar ungeduldig sein.
Die Tatsache, dass npk industrial design diese störrische Leidenschaft seit 25 Jahren mit Erfolg gegenüber seinen Auftraggebern durchzusetzen weiß, zeigt, dass die Ergebnisse überzeugend sind. Auch international gibt es nur wenige Designbüros, in denen so analytisch und systematisch und gleichzeitig so intensiv und leidenschaftlich entworfen wird, wo die Selbstkritik so groß geschrieben und wo ein derartiges Durchhaltevermögen an den Tag gelegt wird. Der Wunsch nach Perfektion und das Bedürfnis, einem Design alles abzuverlangen, veranlasst sogar npk industrial design manchmal, auf eigene Faust und eigenes Risiko, die Spezifikationen eines Auftrags zu überschreiten. Manchmal sind die Grenzen aber schon erreicht und es gibt keinen neuen Schritt mehr. Doch es kommt regelmäßig vor, dass ein Auftraggeber mit einer neuen Idee überrascht wird, die seinem Produkt eine zusätzliche Qualitätsdimension verleihen kann.

Entwicklungen in Tiefe und Breite
Seit der Ausstellung im Stedelijk Museum ist npk industrial design in Tiefe und in Breite gewachsen. Als logische Entwicklung ergaben sich neue Fachgebiete, unter anderem das Grafikdesign. Viele Produkte tragen grafische Elemente wie Markennamen, Markenbilder oder Gebrauchsanweisungen. Bei vielen Produkten verläuft die Interaktion, die Bedienung und Kontrolle zwischen Benutzer und Produkt über ein 'User interface'. Für all diese Aspekte wurde die Abteilung Grafikdesign ins Leben gerufen, die weit über die Unterstützung der Produktdesigner hinausgeht und mittlerweile einer Eigendynamik folgt.

Für Spezialprojekte wird zudem intensiv mit externen Fachleuten zusammengearbeitet, etwa mit dem Typographen Gerard Unger bei der neuen, landesweiten Ausschilderung der ANWB in den Niederlanden.

Durch das persönliche Interesse von einem der Partner für die Gestaltung des öffentlichen Raums ist eine neue Produktlinie mit Bänken, Straßenbeleuchtung und Wegweisersystemen entstanden. Inzwischen hat sich in diesem Bereich so viel Wissen und Erfahrung angesammelt, dass npk industrial design eine völlig neue Ausstattung von Straßenbahnlinien, einschließlich Haltestellen und Beleuchtungssystemen in eigener Verantwortung realisieren kann. Für die neue Hochgeschwindigkeitsstrecke (HSL) entwickelt npk industrial design außer der Ausstattung auch die Spezialtrasse, ein modular, industriell zu produzierendes Produkt aus Beton. Normalerweise hätte man damit ein Ingenieurbüro für Wasser- und Straßenbau beauftragt.

npk industrial design hat seine Flügel international ausgebreitet. Seit einigen Jahren gibt es eine Dependance in Hamburg, um den zahlreichen deutschen Auftraggebern in logistischer und kultureller Hinsicht näher zu sein. Eine Dependance in Boston (USA) ist im Aufbau.

Neben der Verantwortung eines erfolgreichen Büros verrät dieses Buch auch die emotionale Bindung von npk industrial design mit dem Medium Fotografie, das weit mehr beinhaltet als die bloße Produktdokumentation. Von Anfang an zeugen das eigene, professionelle Fotostudio und die Zusammenarbeit mit externen Fotografen von diesem Interesse des Büros. Zu nennen wäre Gerrit Schreurs, dessen Fotos nicht lediglich ein Abbild seiner Wahrnehmung der Produkte darstellen, sondern vielmehr das Ergebnis eines eingehenden Dialogs mit dem Büro. Auf den ersten Blick scheinen sie wie eine erste Begegnung mit npk industrial design selbst, sachlich und zurückhaltend. Bei näherer Betrachtung jedoch bemerkt man, wie die persönlichen Intentionen des Designers durch den Ausdruck des Materials und den raffinierten Einsatz von Schatteneffekten akzentuiert werden. Für die Fotografie im eigenen Studio wurde gemeinsam mit Gerrit Schreurs eine Bildform entwickelt, mit der npk industrial design die Bildpräsentationen einheitlich gestaltet.

Struktur und Intention Vision & Precision

Vision & Precision gibt überraschend und direkt Einsicht in tiefer liegende emotionale Überlegungen, die ein Design bestimmen. Zwei Probleme werden elegant gelöst, die immer dann auftreten, wenn wir uns Abbildungen von Gebrauchsgegenständen ansehen. Eine Produktabbildung ist quasi die nackte Darstellung der Wirklichkeit. Einerseits stimmen die Maße nicht, andererseits stört die Perspektive im abgebildeten, zweidimensionalen Bild die Einschätzung der Größe. Die Rückseite ist nicht sichtbar, jede Bewegung fehlt, während viele unserer Gebrauchsgegenstände sich ja gerade bewegen müssen, um ihre Funktion erfüllen zu können.

Die Wahrnehmung und dementsprechend unser Urteil von vielen dieser Produkte, ob Kindersitz, Stuhl oder Laborinstrument, wird in bedeutendem Maße durch die taktilen Erfahrungen bestimmt und diese fehlen beim Betrachten einer Abbildung. Aspekte wie Gewicht und Balance, Textur und Temperatur der Oberfläche, Widerstand und Geräusch von Schaltern und Griffen und die wechselnden Lichtreflexionen an der Oberfläche sind nicht wahrnehmbar. Was aber vor allem fehlt, ist der Kontext, in dem die abgebildeten Produkte funktionieren, denn immer stehen unsere Gebrauchsgegenstände in Wechselbeziehung zu anderen Objekten.

Bei der Betrachtung von Produktabbildungen gibt es ein weiteres Problem. Unsere industriellen Gebrauchsgegenstände spielen im 20. Jahrhundert erstmals eine Doppelrolle. Das Produkt sollte sowohl seiner Funktion entsprechen, als auch Träger der persönlichen Identität des Besitzers sein. Es kennzeichnet sich durch nüchterne und treue Funktionalität im täglichen Leben und dient damit unserem Komfort und unserer Sicherheit bei der Verrichtung von Aufgaben. Daneben hat die Formgebung eine bemerkenswerte und beherrschende Rolle in der nichtmateriellen

Expression unseres Selbstbildes bekommen, oder des Bildes, das wir gern darstellen würden. Es scheint, als hätte der Benutzer einen Teil seines Selbstbildes an seine materielle Umgebung abgetreten. 'Zeigen Sie mir Ihre Gebrauchsgegenstände und ich sage Ihnen, wer Sie sind', ist mehr als eine lapidare Bemerkung. Offensichtlich können demselben Produkt verschiedene individuell bestimmte Interpretationen angehängt werden. Der Gebrauchsgegenstand avanciert damit zum willigen Träger einer ganzen Bedeutungspalette.

npk industrial design hat eine überraschende und wirksame Lösung gefunden, beide Wahrnehmungsprobleme für den Benutzer dieses Buches sichtbar und plausibel zu machen.
Neben jeder Abbildung eines Gegenstandes steht als untrennbares Pendant eine weitere Abbildung. Die Produkte werden isoliert und ohne Kontext gezeigt, dem Pendant wird die eigentliche Aufgabe zuteil. Die Bilder - meistens Fotos - rufen beim Betrachter spontan allerlei Interpretationen hervor, mit denen seine persönliche Wahrnehmung von der Abbildung des Produkts beeinflusst wird. Diese Pendantbilder verdeutlichen direkt, dass das Individuum auf eigene Weise, mit eigenem Wissen, eigener Einsicht und vor allem mit seiner persönlichen Erlebniswelt die Gegenstände um sich herum betrachtet. Deutlich wird auch, wie divers unsere Gebrauchsgegenstände interpretiert werden können und wie wenig der Designer hier eingreifen kann. Er kann, da es keine eindeutige Wahrnehmung gibt, nicht steuern, wie der Benutzer das Design zu sehen hat. Seine Stärke als Designer muss darin bestehen, eine Form anzubieten, die eine mehrdeutige Interpretation fördert, ohne an Persönlichkeit zu verlieren.
Der Einsatz von Pendantbildern hebt zum Teil auch das genannte Problem der unvollständigen Darstellung der Wirklichkeit auf. Dieses Phänomen ist heutzutage nicht mehr so störend, weil die spontanen Assoziationen eine als virtuell zu beschreibende Realität aufrufen, in der das unvollständige Bild eine eigene abgeleitete Wirklichkeit aufruft, die überdies einen emotionalen Inhalt bekommt. Das Buch verdeutlicht also nicht nur, womit sich npk industrial design beschäftigt. Auf visuelle, anregende Weise werden dem Betrachter zudem die vielen, nuancenreichen Beziehungen, die wir zu hunderten verschiedener funktioneller Produkte im täglichen Leben haben, bewusst gemacht.
Vision & Precision ist bestimmt kein Bilderbuch. Es entsteht ein Spannungsfeld zwischen Bild und Betrachter. Dieser begreift, dass seine Assoziationen in einem Monat oder einem Jahr anderer Art sein werden, da er täglich Neues erlebt und täglich Neues lernt.

Reyer Kras

Reyer Kras (1940) studierte Industrial Design und Kunstgeschichte. Er unterrichtete an der Universität in Delft (Industrial Design) und an der Reichsuniversität in Leiden (Design Geschichte). Von 1982 bis 2001 war er als Kurator Industriedesign für das Stedelijk Museum in Amsterdam tätig. 1996 gründete er das Büro Aurelius Industrial Design für Produktentwicklung und Ausstellungsgestaltung. Er hat viel publiziert und unterrichtet regelmäßig im In- und Ausland.

npk industrial design bv
Noordeinde 2d
2311 CD Leiden
The Netherlands
T +31 (0) 71 514 13 41
F + 31(0) 71 513 04 10
npk@npk.nl
www.npk.nl

npk industrial design GmbH
Mittelweg 158a
20148 Hamburg
Germany
T +49 (0)40 363828
F + 49 (0)40 378007
npk@npkdesign.de
www.npkdesign.de

npk industrial design inc
burlington center
25 burlington mall road
suite 300, burlington,
MA 01803, USA
T + 1 781 270 0654
npk@npkdesign.com
www.npkdesign.com

book concept	Peter Krouwel
cover concept	Simon Wilkinson
graphic design	Dafne Arlman, Esther van Houdt, Patrick Nas, Liza Snook
image selection	Dafne Arlman, Esther van Houdt, Ronald Lintjer
image research	Marjan Mak, Monique Wittkampf
photo captions	Peter Krouwel, Marjan Mak, Wolfram Peters
translation	Jane Bemont, George Hall, Ellis van Ramesdonk
lithography	Chromalith Repro, Leiden

BIS Publishers, Nieuwe Spiegelstraat 36
1017 DG Amsterdam, the Netherlands
T +31 (0)20 620 51 71, F +31 (0)20 627 92 51
bis@bispublishers.nl
www.bispublishers.nl

ISBN 90-6369-035-5

Printed in Singapore.

npk industrial design

Jacqueline van den Boom: 18, 19, 27, 39, 42, 46, 59, 70, 71, 75, 81, 88, 89, 101, 104, 105, 108, 112, 116, 123, 126, 137, 148, 161, 167, 174, 184

Mascha Hendriks: 11, 25, 26, 32, 40, 53, 54, 57, 69, 72, 73, 76, 79, 85, 90, 103, 109, 115, 130, 140, 142, 158, 173

Peter Krouwel: 65, 114
Vincent Nabbe: 113, 157
Henk de Roij: 117
Liza Snook: 94
Peter Wigman: 97

© ADAGP, 2002 c/o Beeldrecht, Hoofddorp (NL): 164
Alcan Technology & Management Ltd., Neuhausen (CH): 56
Van Beek Digital Images: 20
Museum Boerhaave, Leiden (NL): 118
CEPEC/PANORAMIC: 49
Courtesy Sonabend Gallery: 176
Fontshop Benelux: 52
Fotostock: 17, 24, 28, 36, 60, 106, 119, 133, 143, 147, 150, 169, 180
© 2000 Dirk R Frans/ dirk.r.frans@xs4all.nl: 44
Getty Images: 29, 77, 99, 121, 159, 165
© Aart Klein/Nederlands Foto archief: 47
Paul Lagro: 22
Mauritshuis, Royal Cabinet of Paintings, The Hague (NL): 151
Anita Meuleman Fotografie: 110
Naturalis, Museum of Natural History, Leiden (NL): 13, 83,87, 125, 135, 139, 185
Radi Designers: 154
Salto Mortale, thanks to Fokker Heritage Trust: 64,168
© by VAN RIESEN®, Product branding: SCHILLER® BRAND COMPANY: 34
S+B: 63
Schweizerische Beratungsstelle für Unfallverhütung, Bern (CH): 95
Eric Schoevers: 128
© Steelcase: 58
Dirk Pauwels: 68
Pix/Giraudon: 37
Sanne Pleiter: 172
Gerrit Schreurs: 14, 15, 23, 31,38, 50, 51, 55, 62, 67, 74, 78, 100, 122, 127, 131, 141, 144, 145, 149, 153, 156, 161, 166, 170, 175
Spaarnestad Fotoarchief & NFP Photography: 33, 129
Miro Ŝvolík, Prague (CS): 179, 182, 183, 186
© Piet Zwart, Man on woodpile, 2002 c/o Beeldrecht, Hoofddorp (NL): 138

1978	Founding of Ingenieursbureau Peters en Krouwel by Wolfram Peters (1952) and Peter Krouwel (1952) in Delft.
1985	Bruno Ninaber van Eyben (1950) joined Ingenieursbureau Peters en Krouwel. The name changed to Ninaber|Peters|Krouwel, industrial design. 7 employees
1990	Relocated to Witte Poort Kazerne (listed building), Leiden. 25 employees
1991	'Function Forms Style' catalogue and exhibition in Stedelijk Museum, Amsterdam.
1996	Founding of npk industrial design GmbH in Hamburg, Germany.
1997	Bruno Ninaber van Eyben left the firm. The name changed to npk industrial design.
1998	Jan Witte (1966) and Jos Oberdorf (1962) became partners.
2001	Founding of npk industrial design Inc. Boston, USA.
2002	Detlef Rhein became partner in npk industrial design GmbH.
2003	Total of 60 employees: designers, engineers, model makers.

npk industrial design has received over one hundred international design awards and honourable mentions.
Visit www.npk.nl for more details.